UX Optimization

Combining Behavioral UX and Usability Testing Data to Optimize Websites

W. Craig Tomlin

Apress®

UX Optimization: Combining Behavioral UX and Usability Testing Data to Optimize Websites

W. Craig Tomlin
Cedar Park, Texas, USA

ISBN-13 (pbk): 978-1-4842-3866-0 ISBN-13 (electronic): 978-1-4842-3867-7
https://doi.org/10.1007/978-1-4842-3867-7

Library of Congress Control Number: 2018958918

Managing Director, Apress Media LLC: Welmoed Spahr
Acquisitions Editor: Susan McDermott
Development Editor: Laura Berendson
Coordinating Editor: Rita Fernando

Cover designed by eStudioCalamar

Cover image designed by Freepik (www.freepik.com)

Distributed to the book trade worldwide by Springer Science+Business Media New York, 233 Spring Street, 6th Floor, New York, NY 10013. Phone 1-800-SPRINGER, fax (201) 348-4505, e-mail orders-ny@springer-sbm.com, or visit www.springeronline.com. Apress Media, LLC is a California LLC and the sole member (owner) is Springer Science + Business Media Finance Inc (SSBM Finance Inc). SSBM Finance Inc is a **Delaware** corporation.

For information on translations, please e-mail rights@apress.com, or visit www.apress.com/rights-permissions.

Apress titles may be purchased in bulk for academic, corporate, or promotional use. eBook versions and licenses are also available for most titles. For more information, reference our Print and eBook Bulk Sales web page at www.apress.com/bulk-sales.

Any source code or other supplementary material referenced by the author in this book is available to readers on GitHub via the book's product page, located at www.apress.com/9781484238660. For more detailed information, please visit www.apress.com/source-code.

Printed on acid-free paper

*Dedicated to my wife, Roe, and to my daughter, Kaila,
without whose love, support, and guidance I simply would
not be the man I am today.*

Table of Contents

About the Author

W. Craig Tomlin is a senior user experience strategist, researcher, and marketing conversion optimization expert. He is a Certified Usability Analyst who conducts UX research, CRO, and UX design improvement to increase ROI. He consults with start-ups as well as the Fortune 500 and has been optimizing websites, mobile sites, and apps since the mid-1990s. He has worked with firms including AT&T, BlackBerry, Countrywide Home Loans, CDC. gov, Disney, DirecTV, IBM, Kodak, Marsh & McLennan, Prudential Insurance Company, Sprint, Verizon, WellPoint Health Networks, Zurich Insurance Company, and many more. He's a former President of the User Experience Professional's Association, Austin chapter, and a speaker at conferences including UXPA, SXSW, World Usability Congress, InnoTech, and more.

About the Technical Reviewer

 **Ritvij Gautam** is the co-founder of TryMyUI Inc. He hails from Mumbai, India and came to the United States to study physics and philosophy at Claremont McKenna College. After specializing in the area of Philosophy of Mind and interning at LAM Research's Global Quality department, Ritvij started TryMyUI at the age of 21 with his co-founder, Timothy Rotolo, to help companies ensure that their websites and apps are usable for their target demographic customers. Since then, TryMyUI has grown to be one of the leading user testing platforms in the world with customers like Cisco, National Geographic, TDBank, American Eagle, Optimizely, Aflac, and many more.

Ritvij is also the Global Entrepreneur in Residence at San Jose State University where he mentors student-run start-ups and guest lectures in classes about entrepreneurship, bootstrapping, conversion rate UX, and more. He and Tim co-authored the book *This SUX!* which is available on their TryMyUI website.

In his free time, Ritvij is an avid squash player, rock climber, and coffee drinker.

Acknowledgments

This book, like many things in our lives, did not happen with just one person. It takes a team to write, edit, and publish a book. So before we begin, I would like to take a moment to thank the many people without whom this book simply could not have been made.

First and foremost, my deepest heartfelt thanks to my very patient friends and family and especially my daughters, Kaila and Katie, and my wonderful wife, Roenia. They were an inspiration for me and provided unconditional support, positive energy, and a never-ending drive that is so necessary to help get a writer though those times when a blank page stares back with a defiant "I dare you to write something here" attitude. My mom and dad also deserve my sincerest thanks for all their love, support, and wisdom over the years, without which I could not do what I do today.

I also want to thank my technical reviewer, Ritvij Gautam, the very busy co-founder and CEO of the UX research firm TryMyUI Inc. and a good friend of mine. His patience and keen insights were a blessing in taking this book from good to great. TryMyUI, for those of you who may not be aware, is a leading usability testing platform that provides cutting edge user testing technology to companies like FitBit, National Geographic, and more.

The team at Apress Media was also wonderful to work with and a huge help in getting this book published. Susan McDermott, Laura C. Berendson, and especially Rita Fernando showed why it takes massive amounts of patience and persistence to keep an author moving, even when that author is staring vacuously at a blank, white page that refuses to add words to its emptiness. Many more hard-working people at Apress Media worked diligently to produce this book and I am very grateful for their time, energy, and passion.

And finally, I have to thank the many clients, companies, and professionals I've worked with and met over the years who have all helped me to hone and improve my craft. Without them, I would not be the numbers nerd and UX geek I am today, and I definitely wouldn't have had the experiences needed to create this unique and powerful approach to UX optimization.

Introduction

Thank you for choosing this book from the millions and millions of books available to you. I suspect that not only do you have good taste in books but you also most likely share my keen interest in data, User Experience (UX), and optimization.

I promise you this will be a very interesting, informative, and hopefully entertaining training guide that will explain step by step how to optimize the user experience of websites and apps using behavioral UX data coupled with UX research and usability testing analysis. I'll show you the process and the steps necessary to accomplish the analysis and optimization methodology I use every day.

I'll also be providing you with examples and stories from my experiences that helped me to create this methodology. And I'll be doing all that with a slightly snarky and hopefully entertaining jauntiness that makes reading and using this book informative and fun.

This book is written primarily for two audiences. The first is people I affectionately refer to as *numbers nerds* (you know who you are). Some of my best friends are numbers nerds. They are people who typically use quantitative data to evaluate and optimize anything they may be working on.

You'll no doubt recognize this group; they're the people generally heads-down in huge Excel spreadsheets, putting pivot tables and cross tabs together to crunch big data the way a camper zealously puts together s'mores before delightedly devouring them.

The second group of people I affectionately call *UX or usability geeks* (and you know who you are). I'm a self-confessed member of this group. I also have many friends who fit this group. We are the people who typically use qualitative data for optimization. You may recognize us; we're the people generally annoying friends and family by keenly watching them use a device and sometimes asking them to "try this and tell me what you're thinking while you're doing it." We're also the ones who delight in pointing out usability flaws, such as bad signage on freeways or door handles on doors that actually push in to open.

As to me, my career started in marketing and advertising in the 1980s prior to the dawn of the Internet. The '80s were an awesome time to be in the tech world. Dinosaurs had just recently died out, and we were all reveling in our big hair and padded shoulders at the amazing new technology that allowed us to instantly send documents anywhere at any time—known as the fax machine.

In the mid-1990s I transitioned into the Internet world of marketing and advertising; in the early 2000s I moved into designing and optimizing email and web-based experiences using Human Computer Interaction and Usability (which today is often associated with the term "UX"). I've worked with or for many big and small companies over the years including AT&T, BlackBerry, CDC.gov, Dell, Disney, EMC, IBM, Prudential Insurance Company, Sprint, Verizon, Virgin Media, Vodafone, WellPoint Health Networks, Zurich Insurance Company, and many, many more.

Whether in the 1980s or today, traditional or Internet-based marketing lives and dies by big data. Impressions, Cost Per Thousand Impressions, Number of Pieces Mailed, Number of Responses, Open Rate, Clicks, Click-Through Rate, Average Time on Page, Leads, Sales, and many, many more data points are the world of marketing. With marketing, big data is all about evaluating the behavior a marketing campaign is producing. When applied to the Internet, I eventually came to call this type of engagement information "behavioral UX data."

After becoming a Certified Usability Analyst in the early 2000s I transitioned into designing and optimizing web experiences. The big data numbers were augmented with qualitative data from usability and related UX research. The qualitative data typically comes from usability testing, task flow evaluations, card sorts, eye tracking, heuristic reviews, reverse tree testing, five second tests, and many other qualitative sources. I eventually referred to this data as "UX and usability testing data."

I eventually came to realize that I received the best results for website optimization when I applied both behavioral UX data (the quantitative side) and usability testing data (the qualitative data) together.

I've worked with many teams over the years, and it's been my experience that rarely will quantitative numbers nerds hang out with and use the methods of their qualitative UX geeks brethren. And I've observed that it's also a rarity for the qualitative UX geeks to crunch and use big data the way their quantitative numbers nerd brothers and sisters do.

And that's sad.

I realized that by bringing both quantitative and qualitative data together a much more comprehensive picture of "what's happening" and "why it's happening" activity can be formed for websites or apps. And it's this far more robust and comprehensive set of data that makes for a far more informed, and thus better, approach to optimization. I call this combination of quantitative and qualitative data a "360 degree view" into *WHAT* is happening (quantitative) data and *WHY* it's happening (qualitative) data, which then informs a far more accurate analysis and set of optimization recommendations.

The goal of this book is to provide you with step-by-step instructions to show you the process I use, so you can use it yourself. You'll learn the art and science of combining both sides of the quantitative and qualitative data spectrum to form a much more robust 360 degree view into your website or app activity. This enables you to provide a more accurate set of optimization recommendations.

Using this approach I believe you can become a better optimization professional. And you may just be able to eat s'mores and annoy your friends and family with the very best of them along the way!

CHAPTER 1

UX Optimization Overview

A long time ago, when the Internet was fresh and new and everyone used AOL and almost nobody had even heard of Google, I walked into my supervisor's office to ask for a promotion. I approached him and suggested that because of several successes I had in this brave new "New Media" world perhaps a promotion was in order. After a long, pregnant silence, he slowly looked up from his paperwork (yes, we used paper to do work in those days) and said,

"You don't know what you don't know."

After a rather long and awkward silence in which he simply stared at me with a somewhat pleasant but clearly finished manner, I realized this very Zen of meetings was over and left his office contemplating his onion layer-like words.

I mulled over his observation often as the years passed and the Internet grew up. Oddly enough, after years of trial and error developing and optimizing hundreds and hundreds of big and small websites and then apps, I finally understood his most Zen of thoughts.

UX optimization, I realized, is just like that.

You don't know what you don't know.

Of course, he hadn't meant it specifically about UX optimization. After all, back then there was no "UX." No, his comment was a more overarching statement about information, observation, and the ability to understand the big picture in a broader, more holistic context.

I came to the conclusion that his statement perfectly summarizes how UX optimization is typically carried out today.

You don't know what you don't know.

© W. Craig Tomlin 2018
W. C. Tomlin, *UX Optimization*, https://doi.org/10.1007/978-1-4842-3867-7_1

I've observed over the past years that in general there are two different types of UX optimization, and two different types of optimizers:

- The first type is typically composed of the big data analysts, the A/B or multivariate testers, and related numbers practitioners. They approach optimization from a quantitative data and metrics analysis perspective. Quantitative data can be thought of as the "what's happening" data. Often qualitative practitioners reside in marketing or product management positions within a firm. This group can flex numbers and cross-tab spreadsheets like Arnold Schwarzenegger used to flex his biceps.

- The second group is typically composed of the usability and user experience practitioners. They are generally the UX and usability focused types of people who are interested in qualitative data, like how satisfied a user is, how a user feels about an action, or how easy or difficult it is for the user to accomplish a task. Qualitative data can be thought of as the "why it's happening" data. This group can analyze qualitative data the way Sigmund Freud analyzed his patients to root out angst about their mothers.

Each group is very good at what they do.

The quantitative group can A/B test with the best of them. And the qualitative group can usability test and UX research the hell out of a design.

Yet seldom do both groups come together, and seldom does one person do both quantitative AND qualitative as part of their day-to-day role.

Thus, quantitative practitioners may know the "what's happening" information associated with the quantitative data of website behavior, but they often don't know the qualitative "why it's happening" side.

Likewise, the qualitative practitioners may know a great deal about the "why it's happening" information associated with the qualitative data, but they often don't know the quantitative "what's happening" side.

They don't know what they don't know.

And that is precisely why I wrote this book.

My goal for this book is to help guide you in the ways of combining these two powerful sets of data into a broader, far more robust and holistic context that improves your ability to analyze and optimize websites and apps.

You will be combining the "what's happening" quantitative data with the "why it's happening" qualitative data to enable much more accurate analysis and subsequent recommendations for optimization.

If you fall more into the quantitative big numbers group, you will learn what you need to know to apply qualitative data for analysis. And if you fall more into the qualitative group, you will learn what you need to know to apply quantitative big-numbers data and analysis.

Simply put,

You will know what you didn't know.

The Four UX Optimization Steps

At the 30,000-foot level, there are four steps to combining quantitative and qualitative data for UX optimization. They are the following:

Step 1: Defining Personas

Step 2: Conducting Behavioral UX Data Analysis

Step 3: Conducting UX and Usability Testing

Step 4: Analyzing Results and Making Optimizations

Let's cover each of these steps in a bit more detail.

Step 1: Defining Personas

Defining and using Personas is the first step in any UX optimization process. That's because it is critical for you to know who you are trying to improve the website for. Let's face it: it is hard to admit but not everyone in the world will find your website or app useful or helpful.

Note   In the interest of more efficient reading, I'll refer to "website optimization" for "website and app optimization." Just know that all the methods you will use in this book can be applied equally for websites or apps.

So who out there MAY be interested in your website and the products or services you offer?

More than likely it's someone with a NEED your product or service helps address. It's probably someone who is SEARCHING for a solution you provide. And it's probably someone who has this need at a time that causes them to be searching for this product or service NOW.

And unless you're selling a $300 million luxury island, the odds are your someone is not alone. There are (or at least so your firm hopes) many, many others who all share that NEED, are SEARCHING for the solution, and are doing it NOW.

Guess what? All those someones share several things in common and because of that you can group them all together into a single representation called a "Persona." Figure 1-1 is an example of a typical Persona.

Persona - Jessica

Jessica – Shopper Mom

Education	H.S.
	College
	Advanced Degree
Job situation	**Not employed**
	Part time
	Full time
	Full time Student
Computer type	**Smart Phone**
	Tablet
	Desktop
Computer tools	Advanced applications
	Coding tools
	Email
	Web browsing
	Word processing
Computer skills	Limited
	Moderate
	Advanced
Domain expertise	**Low** Medium High

Shopper Mom:

Jessica is a 45-year-old mom of a son who enjoys dressing well. Her son is familiar with several stylish watch brands. For Christmas, Jessica's son has provided a short wish list of items he would like, including several items from one of the high-end watch brands. Jessica is not familiar with the brand, but needs to use the site to try to find a gift her son chose. Importantly, she's on a budget so price is important. She needs to find a gift her son likes, but at the price limit she has set for herself.

Jessica's critical tasks:

1. Find the items her son indicated he liked.

2. Evaluate the cost for each item.

3. Purchase the item that meets her specific budget.

Figure 1-1. *"Personas" are representations of typical users, based on shared critical tasks*

You need to use that Persona to help you focus on who you are optimizing the website for. You will use the Personas' needs, their searching behaviors, and their mental map for how they typically research and find a solution. You will also use other behavioral elements they share to help you understand how your site is performing in helping them achieve their critical tasks.

You need Personas because you must analyze the critical tasks necessary for them to be successful on your website.

I will go into greater detail later in the book on how to create Personas and how to use them for analysis, in case your firm doesn't use them already. But for the purposes of the four UX optimization steps, let's move on to the next step.

Step 2: Conduct Behavioral UX Data Analysis

Step 2 is to conduct behavioral UX data analysis to evaluate the quantitative data associated with Persona activity on your website.

Now that you know the Persona and what behaviors they have, you can evaluate those behaviors on your website. This data is quantitative because it is the "what is happening" data. You need to analyze the existing user experience of the website based on this quantitative behavioral data.

Your goal is to find and evaluate the quantitative data in the context of understanding how your Personas are, or are not, accomplishing their critical tasks.

Where does this behavioral UX data come from? Typically it's found in your web log file analysis systems such as

- Google Analytics (often called just GA)

- CoreMetrics

- Adobe Analytics Cloud

- Or related types of website analysis tools

What types of behavioral data do you look at? This will vary depending on the Personas, the type of website you have (i.e., eCommerce, B2C, B2B, etc.) and what critical tasks (which come from the Personas) and activities your website visitors are conducting on your site.

Common Types of Behavioral UX Data

In general, the most common types of behavioral data you should evaluate in your audit align with the basic user experience of the site, including

- Conversion Data from ERP & GA Systems

- PPC Keyword Data

- Website Conversion Data

- Website Bounce Rate

- Visits by Browser

- Average Time Spent Per Session and Per Page

- And many others, depending on the website and Persona critical tasks

An example of behavioral UX data can be seen in Figure 1-2, which is a website Sessions by Browser report from Google Analytics.

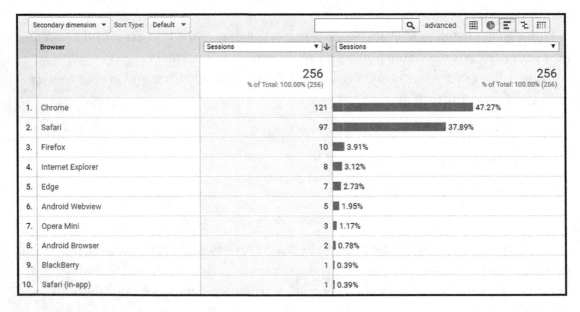

Figure 1-2. *Sessions by Browser Report (Google Analytics)*

So now that you know the types of behavioral UX data you need to audit, you can use that data to have a better sense of "what's happening" on your website.

But that's not enough. So what's missing?

Although you know "what's happening," the behavioral data does not tell you "why it's happening."

For that, you need to switch to finding and using qualitative data analysis (i.e., usability and UX testing).

Step 3: Conduct UX and Usability Testing

You conduct UX and usability testing to help you uncover the WHY of the behaviors you analyzed in the previous step. You do this by observing real people who match your Personas as they try to accomplish their critical tasks on your website.

There are a variety of UX and usability testing tools and data you can use to help you uncover the WHY. The list of what is actually used will vary depending on the Personas, the type of website you have (i.e., eCommerce, B2C, B2B, etc.), and what critical tasks and activities your website visitors are conducting on your site.

Your goals in conducting the UX and usability testing research are to identify

- What parts of the critical tasks work well for your website visitors?

- What parts do not work well for them?

- What confuses or causes them concerns?

- Are their expectations for the experience being met? Why or why not?

Common Types of UX and Usability Testing Data

In general, the most common types of UX and usability data you should evaluate in your audit align with the critical tasks the Personas are trying to accomplish and may include the following:

- Moderated Usability Test

- Unmoderated Usability Test

- 5 Second Test

- Click Test

- Others, depending on the Personas and critical tasks being evaluated

In-person or remote moderated usability testing is generally the richest and most robust way to capture the "why it's happening" data. Figure 1-3 is an example of an in-person moderated usability test in which the screen action is recorded at the same time as the participant's face and voice are being recorded.

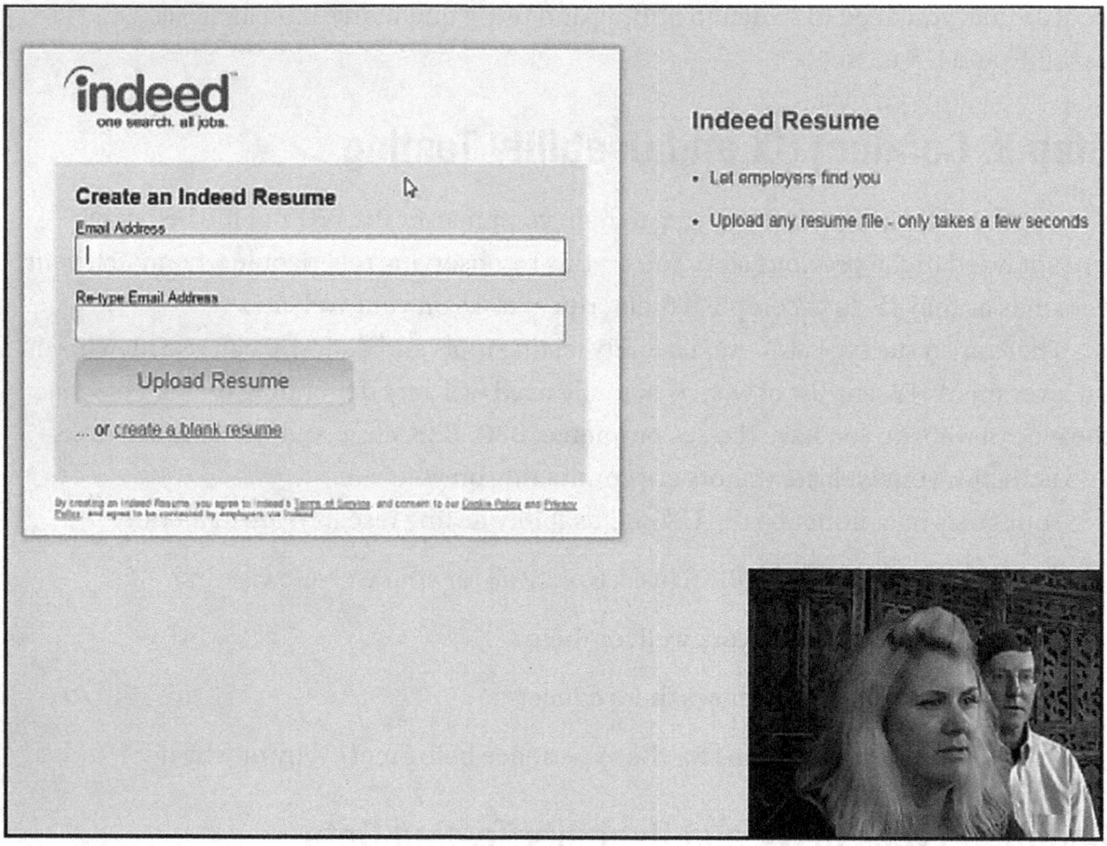

Figure 1-3. *In-person moderated usability test*

There are other good methods for obtaining UX and usability testing data and they include unmoderated usability testing, 5 second testing, question tests, and other types of qualitative UX tests.

The UX and usability testing methods above will provide you with the all-important qualitative "why it's happening" data of the quantitative "what's happening" behavioral UX data you already documented.

Knowing the "what's happening" data, and combining it with the "why it's happening" data, you now have a much clearer picture of the behavior on the site and why that behavior is happening. All that's left now is to analyze that data, combine it into a set of optimization recommendations, and use them in A/B testing of the website.

Step 4: Analyze Results and Make Recommendations

Next, you combine the analysis of behavioral UX data in Step 2 with the UX research and usability testing data from Step 3 to determine the WHAT and WHY for your website interaction.

Your goal is to look for patterns that align with undesirable behaviors. Based on this data, you need to determine where optimization opportunities exist and what changes you believe will improve those behaviors.

The quantitative "what's happening" behavioral data is your signpost; you use it to identify where critical tasks are not performing as expected. You will focus in on those pages or on those parts of the flow that need attention.

The qualitative "why it's happening" data is your tour guide; you use it to identify why those critical tasks are not performing as expected. Often those WHY issues may resolve around one of several common usability issues such as those in the next section.

Common Types of Behavioral UX Issues

In general, the most common types of behavioral data you should evaluate in your audit align with the basic user experience of the site, including

- Taxonomy not in alignment with users

- Navigation errors or confusion

- Process flow not in alignment with user's mental map

- Other heuristic issues depending on the site

Finally, just because the behavioral UX "what's happening" data and the UX research "why it's happening" data seem to provide you with optimization recommendations, you should never assume that your analysis is correct.

My recommendation is that any analysis and set of recommendations always include vetting using A/B testing. A/B testing is the only way to be sure that the optimizations you recommended did in fact actually improve things.

Conclusion: Four UX Optimization Steps

In conclusion, the four big UX optimization steps are the following:

Step 1: Defining Personas

Step 2: Conducting Behavioral UX Data Analysis

Step 3: Conducting UX and Usability Testing

Step 4: Analyzing Results and Making Optimizations

Step 1 is to clearly define who you are optimizing the website for by creating a Persona or Personas, or using existing Personas if they already exist. A Persona is a fictional representation of your most common website visitors who all share the same critical tasks.

Step 2 is to conduct the Behavioral UX Data analysis to identify the quantitative WHAT behavioral data coming from log file analysis tools such as Google Analytics. You can identify potential areas of the website that may be causing poor critical task performance for your Persona or Personas.

Step 3 is to conduct qualitative WHY UX and usability testing to uncover potential reasons for the poor critical task performance. These reasons may align with some of the more common heuristic usability problems that cause website visitors to have difficulty in accomplishing their critical tasks.

Step 4 is where you combine the WHAT behavioral data with the WHY UX and usability testing data. This gives you a much clearer picture of the user experience on the website. With this information, you can now look for patterns and make recommendations for potential optimizations. A/B testing should be conducted on any optimization recommendations to ensure that the recommendations actually did improve the user experience of the website.

CHAPTER 2

What's a Persona?

Step 1 of your UX optimization process is to define a Persona, or use already existing Personas if they are available.

So exactly what is a Persona, and why is it important? For purposes of UX and UX optimization,

> *"A Persona is a representation of the most common users, based on a shared set of critical tasks."*

Defining Personas is the first step in any UX optimization process. That's because it is critical for you to know *who* you are trying to improve the website for. Here's why.

When I slave with loving, tender care over my websites, spending hours and hours making them the best they can possibly be, I grow very, very attached to them. It's kind of like having a baby, but without the messy diapers. Maybe you do the same?

After all that tender, loving, caring, and hard work, it's sometimes difficult to admit that not everyone who happens across your website may like your little bundle of pride and joy. Worse, not everyone may want or need your wonderful website. In fact, there are probably a lot of people in the world who would find your baby, er, I mean your website ugly.

"OMG! For realz?!?"

Like me, you may have a tendency to get so in the details of your site, so familiar with every cute dimple, that you forget a golden rule:

> *Not everyone in the world will use your website, or find it useful or helpful.*

So when you design a new site, or optimize your existing site, you need to remember to understand the perspective, needs, and mental map of the people who actually need to use your website. Your goal is to put yourself in their shoes. You must evaluate what they will find useful and helpful, because otherwise you may be making their life difficult, and they may end up leaving your website forever.

© W. Craig Tomlin 2018
W. C. Tomlin, *UX Optimization*, https://doi.org/10.1007/978-1-4842-3867-7_2

So who are these people that you must focus on?

- More than likely it's someone with a ***need*** your firm helps address.

- It's probably someone who is ***searching*** for a solution you provide.

- And it's probably someone who has this need at a time that causes them to be searching for this product or service ***now***.

And the odds are your someone is not alone. There are plenty of others (or at least so you hope) who all share that need, are searching for the solution, and are doing it now.

And that's good news, because all those someones share certain attributes in common, and because of that you can group them together into a single, fictional representation called a *Persona*.

You need to use that Persona to help you focus on who you are optimizing the website for. You will use that Persona's needs, their searching behaviors, and their mental map for how they typically think about the process of completing their critical tasks. You will also use other behavioral elements they have in common to help you understand how your site is performing in helping them, or not, with those critical tasks.

You will analyze the Persona's common critical tasks that must be accomplished for them to be successful on your website. So clearly, starting with an explicit Persona is the first mission-critical step for UX optimization. I will cover in detail how to create a Persona in a later chapter. But first, a little Persona history is in order.

A Brief History of the Persona

In 1998, Alan Cooper published his ground-breaking book on interaction design: *The Inmates Are Running the Asylum*.[1]

In his book, Cooper describes what Personas are and why he created them, their purpose specific to design interaction, and the traits that make up a Persona.

To summarize Cooper, he created Personas as a way of synthesizing into a single representative user a common set of needs, backgrounds, and knowledge (also known as domain expertise) to define what a "typical" user would need to do to be successful with an application. He gave this Persona a name and even provided a bit of a story about who the Persona was, what their goal or goals were, and how and why they needed to use the application.

---

[1]Alan Cooper, *The Inmates Are Running the Asylum* (Indianapolis, IN: Sams, 2004).

As the years went by, Cooper further refined his Persona. His book also drove many others in design, website development, and interaction design to create their own versions of Personas.

Common Attributes of Personas

Although there's no one "true" formula for a good Persona, the best ones share certain elements, such as the following:

- They are based on actual field research (observing and asking questions of potential users).

- They identify common patterns of behavior.

- They focus on the now, not a potential future state of how things might be.

- They include a picture, name, and brief story to humanize the Persona.

- They describe a problem or task the Persona is trying to solve, typically in a story format.

- Good Personas include the typical environment and/or devices used.

- They include details on the domain expertise of the typical user (how familiar the user is with the language, process flow, and details of usage).

- They call out in prioritized order the top two or three critical tasks the Persona must do to be successful.

According to Cooper, a good Persona should enable you to answer any design question when pondering a function or feature of an application. By using a Persona, it's easier for design, development, and QA teams to align their thinking to the user's critical tasks. This is because the Persona focuses attention away from thinking about features, and instead focuses attention on the end user and what THEY need to be successful. This is called "user-centered design" and Personas are a wonderful design tool when using user-centered design to create applications.

Although there may have been other earlier versions of representative users that could loosely be described as Personas as we think of them today, Cooper's book was instrumental in the rapid adoption and promotion of using Personas in interaction design across the globe.

Types of Personas

In the next chapter, I will provide much more detail about the types of Personas, but as an overview let's cover the basics here.

If you search the Internet for "UX Personas" and review the images results pages, you will see thousands and thousands of varying versions and types of Personas, of which no two seem to be identical. Figure 2-1 demonstrates the image results from a typical Google search for "UX Personas."

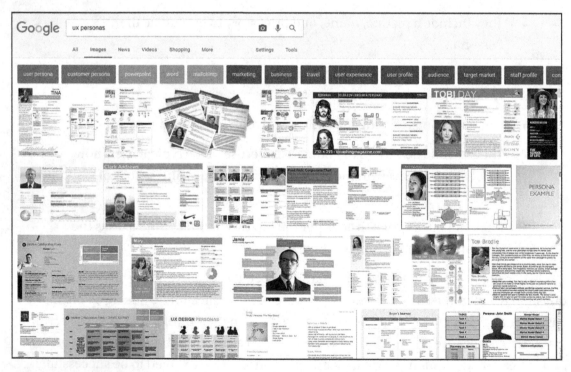

Figure 2-1. *Google image results page for UX Personas*

This can be very confusing, but there are only two general types of Personas (and one "almost" type):

- Design Personas (UX)

- Marketing Personas (Buyer Personas)

- Proto-Personas (made without field research)

Design Personas

The first type of Persona is the Design Persona. Design Personas are most commonly used for UX or Usability task-based work. This type of Persona will normally have critical tasks identified. Figure 2-2 is an example of a typical Design Persona. Note the critical tasks that are identified specifically for assisting with usability testing.

Persona - Jessica

Jessica – Shopper Mom

Education	H.S.
	College
	Advanced Degree
Job situation	**Not employed**
	Part time
	Full time
	Full time Student
Computer type	**Smart Phone**
	Tablet
	Desktop
Computer tools	Advanced applications
	Coding tools
	Email
	Web browsing
	Word processing
Computer skills	Limited
	Moderate
	Advanced
Domain expertise	**Low** Medium High

Shopper Mom:

Jessica is a 45-year-old mom of a son who enjoys dressing well. Her son is familiar with several stylish watch brands. For Christmas, Jessica's son has provided a short wish list of items he would like, including several items from one of the high-end watch brands. Jessica is not familiar with the brand, but needs to use the site to try to find a gift her son chose. Importantly, she's on a budget so price is important. She needs to find a gift her son likes, but at the price limit she has set for herself.

Jessica's critical tasks:

1. Find the items her son indicated he liked.

2. Evaluate the cost for each item.

3. Purchase the item that meets her specific budget.

Figure 2-2. *A typical UX Design Persona*

Marketing Personas

The second type of Persona is a Marketing or Product Development Persona. Marketing Personas are often called "Buyer Personas" and are similar to Design Personas. Figure 2-3 is an example of a typical Marketing Persona. The main difference between Marketing and Design Personas is the fact that Marketing Personas typically do not include critical tasks in the description. A Persona in this group may also be based more on demographic and related aggregated data and somewhat less on actual field research and/or observation of actual users.

Persona – Multitask Mark

Motivator: Do the work for me

Primary Concern	Identifiers	Needs/Wants	Assets

• Mark is an owner-operator who does everything from billing to answering the phone to installing software patches to generating sales.
• Mark juggles many things at the same time and acts like a know-it-all.
• He wants to offer more managed services but is too overwhelmed with break-fix tasks that fill up his day and generate low margins.
• He wants to increase revenue but there are few easy paths to get there.
	• Mark is frenetic and dismissive.
	• For Mark, things are very black and white: complete his ever-growing checklist of activities and dream of a better tomorrow.
		• He needs vendor-partners that can do much of the work for him.
		• He wants simple and automated software, with lots of customizable marketing materials to help him sell it to clients.
		• He wants a partner who is patient because results and scale won't happen overnight.
			• High Funnel:
			  • Email BU/Archive Video
			  • Profit from SaaS Email BU article
			  • Do My Work article
			  • Case Studies
			  • Email Backup and Archiving WP
			  • Email Archiving Infographic
			  • Perception Infographic
			• Mid Funnel:
			  • Email BU/Archive Data Sheet
			  • Perception WP
			  • Example marketing materials
			• Low Funnel:
			  • Services lists with cost/benefits laid out

Figure 2-3. *Example of a Marketing Persona*

Proto-Personas

The third type of Persona is a Proto-Persona. Proto-Personas are similar to Design Personas with the exception that they are made without field research or actual observation of users.

As an example, I worked with several firms whose product teams created Personas based on secondary research and customer service information. They examined sales conversations with prospects and customers. They also referenced customer service calls and spoke with product designers who had documented the competitive space. This data is helpful, but cannot replace actual field research of users. Without that all-important and critical field research and observation (also known as "contextual inquiry") the Persona is not a true Design Persona.

Why Personas Matter

So why do Personas matter? I'll go into more detail in an upcoming chapter, but the main reason why Personas matter is the following:

Personas help focus design and optimization efforts squarely on the user and their needs, reducing any opinion-based or subjective decisions about the design, functionality, or features.

The biggest mistake UX teams make when designing, testing, or optimizing the user experience of a website or application is NOT having a carefully defined Design Persona or set of Personas. Without that Persona, any UX research or testing will almost always be flawed to some degree.

There are three main reasons why UX Personas are important:

- **Design Decisions**: By referring to a Design Persona, design and product teams can make decisions that are based on the perspective and needs of the end user. By evaluating each decision against the yes/no of whether it helps the Persona achieve their goals, all those decisions become easier. This is also true for UX research in which the Persona is critical for user-focused analysis.

- **Quality Checks**: The Persona is a go-to tool for preemptive quality checks, especially for those using an agile development approach. If the team cannot confidently look at a function or design and agree that it benefits the Persona, then that function or design should be reevaluated immediately.

- **Scope Creep**: Using a Persona can help ensure everyone agrees on the final destination for the product. This results in helping to reduce scope creep or potentially negative changes in the design as the website is developed.

17

Conclusion: What's a Persona?

A Persona is a representation of the most common users, based on a shared set of critical tasks. Personas are based on field research and direct user observation in their environment (also known as contextual inquiry). Personas vary broadly, but most generally share the following attributes:

- They are based on field research and user observation.

- They identify common patterns of behavior.

- They focus on the now, not a potential future state of how things might be.

- They include a picture, name, and brief story to humanize the Persona.

- They describe a problem or task the Persona is trying to solve, typically in a story format.

- Good Personas include the typical environment and/or devices used.

- They include details on the domain expertise of the typical user (how familiar the user is with the language, process flow, and details of usage).

- They call out in prioritized order the top two or three critical tasks the Persona must do to be successful.

There are three general types of Personas:

- Design Personas (UX Personas)

- Marketing Personas (Buyer Personas)

- Proto-Personas (made without field research)

Personas are extremely important to the design and optimization of websites and apps because they help focus teams on the end users' needs and mental maps for process flows. Without Personas, any testing or optimization of a website has the danger of being flawed because the analysis and results may not be focused on helping the end user achieve their goals and critical tasks.

CHAPTER 3

Types of Personas

My friend Robert likes to tell the story of when he realized how mission-critical Personas and UX testing actually was to web design or redesign projects. We were working at a large health insurance company at the time. Robert was leading the team that was responsible for updating a web-based portal people would be using to find and access important insurance information.

Tens of thousands of users would be interacting with this portal on a daily basis, so this was a big project with high management visibility and he was very pleased to be working on it.

Having worked on several projects for the company, Robert knew the lingo pretty much backwards and forwards, and he figured anyone else working there would as well.

Robert and his team labeled several main sections of the website with "Producer" and "Provider." It seemed obvious to him and his team that this meant "Agents and Brokers" and "Doctors, Hospitals, and Other Health Care Providers," respectively. In the health insurance industry these were fairly common terms.

For whatever reason (I like to take the credit when I'm telling his story) a thought kept crossing his mind (insert ghostly-Craig voice here):

"Robert! I am the ghost of usability testing future! You should probably consider usability testing your main navigation labels!"

Ghostly Craig aside, he thought it prudent to quickly test the labels just to make sure all was in order.

He took a printed copy of a wireframe with the navigation labels on it and walked down the hall of the insurance company offices. He fully expected that since most everyone he met was either an employee or worked closely with the firm they should all have no trouble identifying the meaning of the "Producer" and "Provider" labels.

But, as you might guess, that was not the case. He found about half of the employees he asked knew what one term meant, but not the other term. He found roughly a quarter of the employees he spoke with didn't know what either term meant. Only a small handful of the people he asked knew what both terms meant.

© W. Craig Tomlin 2018
W. C. Tomlin, *UX Optimization*, https://doi.org/10.1007/978-1-4842-3867-7_3

Robert had learned several valuable lessons:

- **First**, it's always good to usability test, even when only at the wireframe stage and even using hallway intercepts.

- **Second**, he dodged a potentially significant mistake in labeling the portals.

- **Third**, he had not created nor used Personas to help him and his team design the portal from the users' perspective.

Personas, Robert realized, were very important to create, get right, and use. There are three general types of Personas. They are

- Design Personas

- Marketing Personas

- Proto-Personas

You'll examine each of them in more detail in this chapter.

But first, it's time to put away your books, take out your #2 pencils, and sit up straight, because **it's pop-quiz time!**

Let's say you do create a Persona. Once your Persona is created, how often do you feel that Persona should be refreshed?

a) Once every six months

b) Once a year

c) Once every other year

d) Never

e) It depends on the type of Persona

And the answer is (drum roll please)...

It depends! Depending on the type of Persona, and depending on the detail associated with that Persona, a Persona could be refreshed as much as once a year, or as little as once every two to three years. I'll cover the types of Personas and how often they should be refreshed in more detail in this chapter.

In general, it may be a good idea to refresh the Persona more often (say once every six months to a year). But for some types of Personas maybe once every year or every other year is fine for a refresh. And for a very select few Personas, you may not need to refresh more often than say every two to three years or so. The only truly wrong choice in how often to refresh a Persona is to choose "never."

The reason why "never" is a wrong answer has to do with the fact that technology continually changes. And sometimes people change too (especially in the adoption of new technology or new processes). Finally, due to the ever-increasing competitive landscape firms find themselves in, demands on the user experience can also change.

So it becomes clear that Personas need to be updated on an ongoing basis to keep them current with the ever-changing world.

Confusion About Types of Personas

I've seen a fair amount of confusion, and false assumptions, about the different types of Personas. This confusion is equally distributed among all users of the various types of Personas.

For example, confusion can come from marketing or product teams who assume that their marketing Personas (often referred to as "Buyer Personas") can be used for Design Personas. Likewise, design teams may feel their Design Personas are fine for marketing needs. Finally, both groups may confuse the Proto-Persona for Marketing or Design Personas (they are not).

Other misconceptions about Personas include

- False assumptions about the kinds of information necessary for each type of Persona

- Misunderstandings or confusion about how each type of Persona is created

- Assumption that field research and/or contextual inquiry is not needed for Personas (it is)

- Inability to determine how well a Persona type accurately reflects the user group it is designed to reflect

- Lack of understanding of the need to refresh Personas and confirm that the Persona type still accurately represents the user group

- Ineffective or inefficient internal communication about the fact that Personas exist, or how to use the Persona type

- Assumption by teams that they do not need to refer back to the Persona as time goes by and iterations of designs are developed

For these reasons and more, companies often fail to fully utilize Personas. Many companies will only create one or perhaps two types of Personas, without realizing why that is a mistake.

Design Personas

Design Personas (also called UX Personas) are focused on the user, their goals, and the critical tasks associated with helping them achieve those goals. These Personas must be able to help designers, developers, and product teams answer important questions about the usability and user experience of a design. Figure 3-1 is an example of a Design Persona with critical tasks clearly identified.

Persona #2 - Joel

Joel – Cash Advance Junkie

Computer skills	Novice
	Expert
Job situation	Employee
	Manager
	Director
	Owner
Computer type	**Mobile**
	Laptop
	Desktop
Computer tools	**Advanced features**
	Analytics tools
	Email
	Web browsing
	Word processing
Loan Angst	Low
	Medium
	High
Domain expertise	**Medium**

Cash Advance Junkie:

Joel is a 40-year-old owner of a fashionable restaurant. He's having some financial difficulty and has taken a cash advance to provide some working capital. His problem now is the cash advance company takes his credit cards for payment, forcing him back into the cash crunch he originally was in. He's looking for a way out and is researching small business loans. His goal is to find a loan to pay off the existing cash advance, plus provide him enough working capital to get his business out of the cash crunch he's currently in. He wants to know the details because his entire business is now on the line.

Joel's critical tasks:

1. Find a loan to get him out of the cash crunch.

2. Understand exactly how the loan works.

Figure 3-1. *A Design Persona with critical tasks clearly identified*

Design Personas reflect behaviors and the needs and pain points associated with the user experience for that group of users. This is in direct contrast with Marketing Personas, which are reflective of needs and pain points, buying behavior, media consumption, awareness or interest in products or services, and the attitudes and perceptions about brands and products.

The most important thing about Design Personas is they are based on actual field research and observation of real users in their own environment.

Let me repeat that.

The most important thing about Design Personas is they are based on actual field research and observation of real users in their own environment.

Design Personas are narrative: they tell a brief story about the people associated with the Persona and why and how they do the things they do in regards to the website or app.

Good Design Personas are memorable; they will "stick with" a team long after the initial exposure to the Persona. With a good Persona and an engaged team you'll hear conversations using the names of the Personas again and again. A typical design conversation using a Persona might go along the lines of

> *"I think Jane would probably find this functionality easier because it fits in with her mental map for how this part of the process should work."*

Frequently, Design Personas are placed in highly visible places as a way to be continuously present throughout the design and development process. Often teams will create posters of Design Personas and hang them on walls in the office to make them readily available to anyone, and as a reminder to the team of who they are building the website or app for.

As such, Design Personas are important tools for any design and development team. Because of this, Design Personas are worth the time, money, and effort to create them properly and well.

The Role of Design Personas

Design Personas play a variety of roles. Let's briefly examine three of the more common roles: providing context for design teams, focusing on critical tasks, and helping remove influence-based design.

1. **Provides Context for Design Teams**: Design Personas are very effective for providing research insights and information about user goals and pain-points for design and development teams. Because the contextual and environmental information is included with the Persona, teams can refer back to the Persona when making decisions about how, what, and where user experiences can or should occur.

2. **Focuses on Critical Tasks**: Although there are variations in the content of Design Personas, often information about the critical tasks associated with the Persona is included. And that's a good thing, because you need that critical task information to help you when recruiting for UX research and usability testing. Having the Persona's critical tasks clearly documented makes for much better and more focused studies. Besides that, if everyone on the team responsible for design and development understands the critical tasks, it's that much easier to coordinate all efforts around making the best possible experience for that group of users to be successful.

3. **Helps Remove Influence-Based Design**: The next role is a bit of a secret between just you and me. Look over your shoulders; is anyone in earshot? No? Good! You'll have to come close, because I'm going to whisper in your ear about a secret role for Personas that applies especially to larger organizations.

    Are you close? Come a bit closer. [Whispering in your ear] It's a fact that sometimes executives and the C-Suite (CEO, CIO, CTO, CFO, C-whatever-O) will sometimes use their considerable influence, either knowingly or unknowingly, to sway design decisions. I've seen this happen many times in my career. Let's face it; executives didn't get to where they are by not having strong influence. But sometimes their opinion can influence design decisions not based on user-centric needs, but on their own biases or assumptions. Shocking! I know, right?!

So here's the secret. By using a carefully crafted and internally approved Persona you can get the C-whatever-O to accept the fact that what's best for the Persona should be the primary focus of the design. [Whispering very closely in your ear again] Personas are a great way of telling execs that although they love their ideas, it's probably best to leave design to the professional designers—without getting fired!

[Normal voice] OK, that's it for the secret. On with the rest of your exploration of the role of Design Personas.

4. **Enables User-Centered Design**: Finally, and perhaps most importantly, a great role for Design Personas is in helping design, product, and development teams focus on the user as part of a user-centered design methodology. This helps reduce or eliminate the potentially dangerous use of opinions, biases, or past designs when evaluating the best design choices for a website or application. User-centered design is an extremely powerful tool for design, but only if there's a good Persona present to "stand in" and represent those users.

When to Update Design Personas

There is no single perfect time for when a Design Persona should be updated. In general, Design Personas should stay consistent with the goals, needs, behaviors, and mental map of the processes most people use when engaging with your website or app.

There are two very broad categories of types of Design Personas that can be considered when evaluating how often to refresh. They are

- Consumer Design Personas
- Business-Based Design Personas

Consumer Design Personas

Consumer Design Personas are generally consumer focused, although that's not to say they are only associated with buying or purchasing. Anyone who uses a website or app and is not part of the organization that created it might be considered a consumer.

This type of Personas may need to be adjusted every year or every other year, and sometimes more frequently than every year. That's because the consumer space can shift and adjust fairly often, based on many factors, some of which include the following:

- Competitive factors in the marketplace

- Changes in technology or usage of technology

- Shifting consumer needs

- New processes or techniques for completing tasks

- Changes in the mental map for how processes are completed

Business-Based Design Personas

The Business-Based Design Personas category is very broad. It includes anyone who may need to interact with a firm including employees, business partners, vendors, or related types of users. Typically these users require more sophisticated, in-depth systems. Generally their domain expertise is higher than that of the Consumer Design Personas.

Business-Based Design Personas may need to be updated less often than Consumer Design Personas, perhaps once every other year or potentially longer, with some needing more frequent updates. This is not a hard-and-fast rule, because although many businesses are in stable environments, just as many others may be in fluid environments impacted by outside forces that cause frequent shifts in the needs and processes represented by Business-Based Design Personas.

In general, Business-Based Design Personas reflect a somewhat more "stable" environment than do Consumer Personas. This is because businesses often have fixed and formalized approaches to how they manage processes and critical tasks. Also, the people that make up a business typically follow standardized processes and procedures, which often means that what worked for a Business-Based Design Persona last year may work perfectly fine for that Persona this year, and the next.

There are many factors that can impact the timing of Business-Based Design Persona updates. Some of the more common ones include

- New competitive businesses entering the marketplace with new processes or task-flows

- Changes in technology or usage of technology

- Shifting business needs

- New business processes or techniques for completing tasks

- Changes in business users' mental maps for how processes are completed

Marketing Personas

The second type of Personas is the Marketing Persona, sometimes referred to as the Buyer Persona. Figure 3-2 provides an example of a template used by marketing teams when developing Marketing or Buyer Personas.

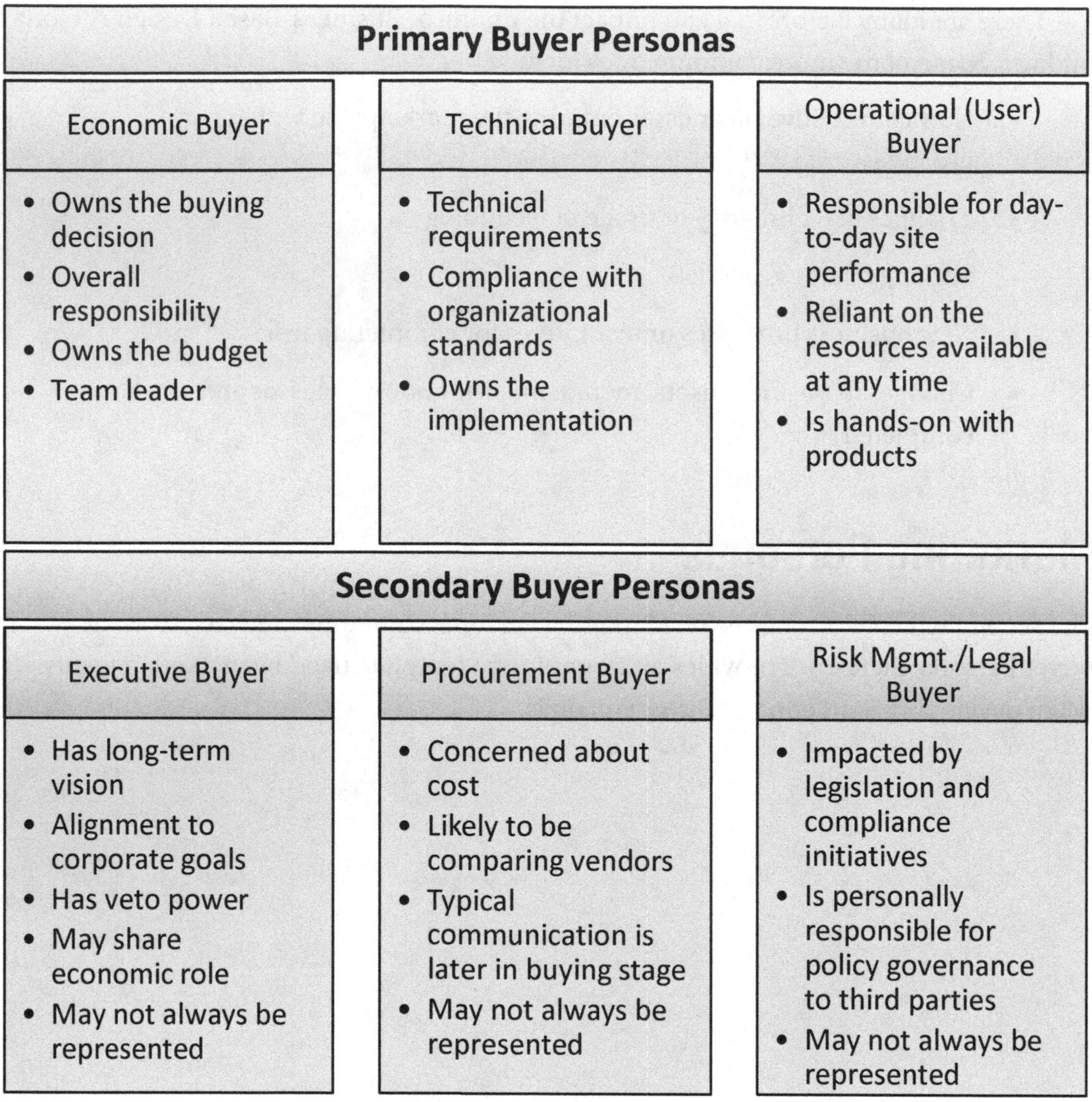

Primary Buyer Personas		
Economic Buyer	Technical Buyer	Operational (User) Buyer
• Owns the buying decision • Overall responsibility • Owns the budget • Team leader	• Technical requirements • Compliance with organizational standards • Owns the implementation	• Responsible for day-to-day site performance • Reliant on the resources available at any time • Is hands-on with products
Secondary Buyer Personas		
Executive Buyer	Procurement Buyer	Risk Mgmt./Legal Buyer
• Has long-term vision • Alignment to corporate goals • Has veto power • May share economic role • May not always be represented	• Concerned about cost • Likely to be comparing vendors • Typical communication is later in buying stage • May not always be represented	• Impacted by legislation and compliance initiatives • Is personally responsible for policy governance to third parties • May not always be represented

Figure 3-2. *An example of a template used for creating Buyer Personas*

Marketing or Buyer Personas are somewhat similar to Design Personas but with a few very important exceptions:

- Marketing Personas generally do not include critical tasks.

- Marketing Personas are often created by using demographic data only, without the use of field research observation (contextual inquiry).

- Marketing Personas often rely on very large sets of quantitative data based on primary or secondary research, although sometimes (sadly) the opposite is true and they are based more on opinions of internal stakeholders than on actual data.

The Role of Marketing Personas

There are several critical roles for Marketing or Buyer Personas in an organization. In fact, most organizations worth their salt will have spent a fair amount of time, energy, and effort in creating and using Marketing or Buyer Personas.

1. **Provides Buying Motivations**: Marketing Personas typically are composed of demographic information and often feature buying motivations and buying needs and desires. This role helps marketing and product marketing teams pinpoint critical messaging as part of a communication plan.

   It should come as no surprise that because Marketing is focused on finding buyers of a product or service, Marketing Personas often have a heavy emphasis on purchasing behavior, purchase intent, and the purchasing motivations involved with shopping for and buying products or services. This data, although critical to good marketing communications, is not as relevant for product design and development needs.

2. **Focuses on Buyer Preferences and Attitudes**: Marketing Personas also often include buyer preferences, opinions, attitudes, media consumption habits, and brand awareness factors. As such, Marketing Personas are used to help explain consumer behavior and consumer perceptions of messaging more than specific processes, task-flows, and critical tasks.

3. **Provides Contextual Clues for Targeted Marketing Communications**: A good Marketing Persona will help marketing and advertising teams create targeted marketing communications and messaging in order to attract attention, stimulate a response, influence shopping and buying behavior, and hopefully result in a sale. Marketing Personas therefore are not as useful for defining what the product is, how the product works, and how and when the product is used.

Because of the importance placed on Marketing (a.k.a. Buyer) Personas in an organization, it's common to see them used in many places besides marketing or product marketing departments. But just because marketing or product marketing teams heavily rely on and use their Marketing Personas often, they're not necessarily the best tool for design.

I've been asked many times about if it is appropriate to use a Marketing Persona for design or UX research work. My answer is … It depends.

If the Marketing or Buyer Persona was developed using actual field research and observation of real users, that's a very good sign you might be able to use that Persona for design or UX research work.

Likewise, if the Marketing or Buyer Persona has critical tasks defined, that is also a good sign you may be able to use the Persona for design or UX research work.

Finally, if the Marketing or Buyer Persona is based on real research and actual user data (and observation!), and not just on assumptions or opinions, that is also a good sign you may be able to use that Persona for design or UX research work.

But don't get your hopes up too much.

I've created, worked with, or consulted on hundreds of big and small Persona creation projects over the past 20 plus years, and in all that time I've only seen maybe a handful of Marketing or Buyer Personas that fit the requirements for use in design work. Most often, the Marketing Personas I've seen don't fit the needs of a UX design or research team. But that doesn't mean the Marketing Personas do not need to be updated. They do!

When to Update Marketing Personas

As with Design Personas, there are many factors that can impact the timing of when a Marketing Persona should be updated. Some of the more common factors include

- New or modified business processes or approaches to work

- Changes in technology or usage of technology

- Shifting awareness, pain points, needs, goals, or priorities

- New approaches, tools, or methods of accomplishing tasks or completing goals

- Adjustments in the mental maps for how processes should be or are completed

Proto-Personas

Of the three general types of Personas in use today, I've found Proto-Personas to be the most plentiful. I believe the reason is because Proto—Personas are the go-to Persona when there is little in the way of money or time to create true research—based Personas. Figure 3-3 is an example Proto-Persona for a health app.

TARGET PERSONALS

Young Families

Young Professionals

Urban Actives

New Movers

Young families looking for a solution especially focused on their children may need the app several times a month.

Young professionals need occasional help thus, may need the app once a month or less.

Urban actives focus on training/health or injury related needs thus, use the app only on an ad-hoc, or limited basis.

New movers focus on obtaining services quickly as they settle in, then use on an ad-hoc basis.

Figure 3-3. *A Proto-Persona for a health app*

Often, Proto-Personas are based on secondary research, and sometimes they are based on little more than a team's best guesses regarding who they are designing a website or application for.

It's a pretty common assumption that using a Proto—Persona to create and test designs is still better than having no Persona at all. And although there's a certain truth to that, nothing beats going out and conducting real contextual inquiry and observation of users in their environment to improve, validate, or optimize a Proto-Persona.

The Role of Proto-Personas

There are several important roles for Proto-Personas, but perhaps the most important of all is their use in agile design and development.

1. **Proto-Personas for Agile Processes**: When faced with an agile process and weekly sprints, most traditional UX research methods for creating and validating Personas are extremely time-challenged in a major way! Just like trying to push your car up a hill in park, trying to create a robust, data-driven, and observation-derived Persona in a week for an agile project is darn difficult!

   Often this means Proto-Personas are used to start the project. Later, validation by way of field research and contextual observation will help "true-up" the Persona. This allows the agile process to continue. I've noticed that many start-up firms seem to favor this approach. Teams can get agile projects underway with a Proto-Persona, and as time goes by conduct field research to transition the Persona into a true Design Persona.

2. **Proto-Personas for Rapid Investigation**: The beauty of a Proto-Persona is it enables very rapid iteration and testing of more permanent Personas. Sometimes it's better to piece together several Proto-Personas with the limited UX research data available than it is to await more robust research. By testing and validating a series of rapidly created Proto-Personas, a team can quickly validate and optimize a more detailed Design Persona using additional, more detailed research methods. As any author will tell you (yours truly included), it's much easier to edit a page already filled with text than it is to create and edit one from scratch.

3. **Proto-Personas as the "Light Beer" of Personas**: As mentioned, the commonly held belief that any Persona is better than no Persona means you must have some sort of Persona for design work. Proto-Personas fill that need. Again, they should not be relied on for long-term projects, but getting something up and

running quickly, based on a limited Proto-Persona can sometimes mean the difference between a design project that succeeds and one that is doomed for failure.

When to Update Proto-Personas

Because Proto-Personas are easy to make, they are also easy to update. I have worked with firms that actively update Personas with each weekly Sprint. This may seem like a lot. But if you're building your Proto-Persona as you go along with each sprint, then updating your Proto-Persona with the key learnings and observations you have from interactions with your design is a good way to continually optimize your Proto-Persona.

But that's the trap with this type of Persona.

The secret with Proto-Persons is to NOT update them with newer versions of the Proto-Persona, as odd as that sounds. Instead, take the opportunity to research real people in their environment and turn those Proto-Personas into Design Personas. The Design Personas, because they are based on real people in their own environment, will prove to be superior to any Proto-Persona made from indirect observations, data, and intuition.

Conclusion: Types of Personas

There are three general types of Personas:

- Design Personas

- Marketing Personas (also called Buyer Personas)

- Proto-Personas

Design Personas are the most useful for website and app design and optimization. That's because they are created by observing real people in their own environment interacting with tasks similar to the critical tasks being researched for the website or app.

There is no single perfect time to update a Design Persona. But they should be updated. In general, Design Personas should be updated as often as necessary to stay current with the goals, needs, behaviors, and mental maps of the people who will be using your website or app.

There are two very broad categories of types of Design Personas that can be considered when evaluating how often to refresh:

- **Consumer Design Personas**: People associated with consumers and individuals using a product or service that are not directly working for or with a firm and who typically have lower domain expertise.

- **Business-Based Design Personas**: People who are either employees or business partners with a firm and who may have a higher level of domain expertise.

Consumer Design Personas can generally be updated every year to every other year, depending on many variables including the product, industry, types of users, and more.

Business-Based Design Personas can generally be updated every year to every two years, or perhaps longer. This is because in general internally-focused applications and related critical tasks the Business-Based Design Persona will be using may not be updated as often as consumer-based applications.

Marketing Personas (sometimes referred to as Buyer Personas) are based on demographic, geographic, and other data sources and typically do not include actual observation of real people in their environment. These Personas are often used by marketing, advertising, and product teams when creating communications for a target audience. The Marketing Personas are most useful for identifying needs, pain points, and desires of prospective customers. This helps marketing teams create the messaging used to attract awareness and engagement of the prospective customers.

Marketing or Buyer Personas, like Design Personas, do not have a set time for when they need to be updated. Instead, various factors can influence when a Marketing Persona should be updated, including

- New or modified business processes or approaches to work

- Changes in technology or usage of technology

- Shifting awareness, pain points, needs, goals, or priorities

- New approaches, tools, or methods of accomplishing tasks or completing goals

- Adjustments in the mental maps for how processes should be or are completed

Proto-Personas are similar to Design Personas with the major exception being Proto-Personas are not created by observing real people in their own environment. Often, secondary sources of data are used to build Proto-Personas, which enables them to be created quickly; this is a useful feature for agile-based firms that require weekly sprints and design research sessions.

Proto-Personas are easy to update because much of the "creation" of a Proto-Persona is based on indirect data and information versus the more difficult and time-consuming work required in finding and observing real people in their environment. Some teams update their Proto-Personas with each sprint using data gathered in usability testing to optimize the Persona. However, this makes Proto-Personas dangerous because, without the direct observation of real users in their own environments, it is possible to arrive at the wrong conclusions regarding why and how users may actually interact with a website or app.

Proto-Personas can and should be updated to true Design Personas at the earliest opportunity.

Now that you know the types of Personas, it's time to turn your attention to the most important question: why Personas matter. You'll explore the answer in the next chapter.

CHAPTER 4

Why Personas Matter

So by now, you may be asking,

> *"Why all this fuss over Personas? Can't we just get on with the metrics and analysis and the optimization and all that good stuff? Who cares about Personas? Why do they matter?"*

Personas do matter, and they should matter so much to you that you wouldn't dream of conducting behavioral data analysis, usability testing, or any UX design project without first having one. Here's why.

Like film in a movie projector, air in your tires, or credit cards at Christmas time, Personas are critical to the UX process. There are many reasons why Personas are so critical, but several that stand out are that they

- Add context to UX behavioral data.

- Enable user-centered design.

- Aid in recruiting for usability testing.

- Decrease scope creep.

Let's examine each of these reasons briefly to better understand why Personas are the start of any good UX project, including UX analysis.

Personas Add Context to UX Behavioral Data

As you will see in the following chapters, there is a LOT of UX behavioral data available. Which parts of that data are useful depends in great part on the Personas. How the Personas should be interacting with the site, versus how they actually are interacting, is an important part of determining what to measure. By using Personas you can better identify what data you should be evaluating in the seemingly ocean-sized amount of data available to you.

© W. Craig Tomlin 2018
W. C. Tomlin, *UX Optimization*, https://doi.org/10.1007/978-1-4842-3867-7_4

Like a kid in a candy shop, some data analysis practitioners may go a bit crazy when looking at all the data available. Tools like Google Analytics seem to be updated annually with even more detailed and complex ways of finding, analyzing, and displaying data. Figure 4-1 provides an example of the tip of the iceberg of data available in Google Analytics.

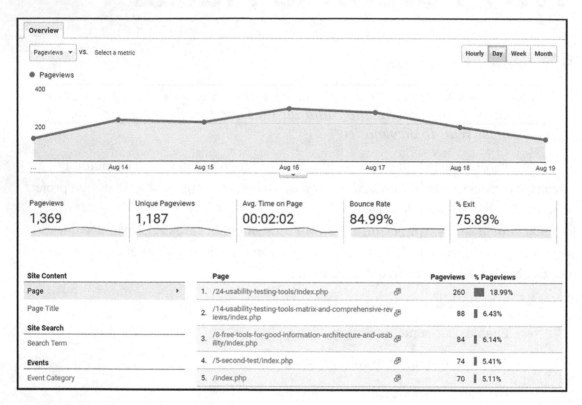

Figure 4-1. *Example of data available in Google Analytics*

So how do you know what information in that sea of data is critical versus just nice to have? Personas will help you identify what data you should be focusing on vs. what's just nice-to-have data vs. data that's not actually usable at all. In a sense, the Persona acts as a filter to help you narrow your focus to the data that actually matters and can help you identify behavioral UX activity.

In the next several chapters, you'll learn how to use Personas and their needs and critical tasks to help find the pertinent data for UX behavioral analysis.

Simply put, without Personas, determining what information is important in UX behavioral data analysis would be almost no better than guessing.

Enabling User-Centered Design

Decisions, decisions, decisions! It takes thousands and thousands of user experience design decisions to create websites, apps, and other software. So when faced with making a choice, how do you know which choice is the best or the most correct decision for your users? Although intuition is helpful (some may argue that to a certain extent we are all "users") it's far better to base decisions by asking

> *"Which way is the best way for the people who will be using this website or app?"*

User-centered design is more than just a catch-phrase; it means including real people in the design process who match the typical user of the system. A great way to do this is by using Personas to help with the many, many design decisions that have to be made each and every day as a new website or app is being developed.

By evaluating each decision against the yes/no of whether it helps the Persona achieve their goals, all those decisions become easier and more accurate in terms of providing real value for the intended users.

A Persona User-Centered Design Story

Not too long ago, I was helping an agile team create an updated version of a form to help people sign up for job interviews. The design team only had a couple of days to come up with a solution for how to design a series of interactions and screens that would allow a person to do the following:

- View a calendar of available interview days.

- Choose an available time for the interview.

- Enter their email to receive a confirmation and reminder.

- Add their scheduled interview to their personal calendar.

The design team was very busy contemplating several design choices for how the interactions would work when I walked into their area. The team members were evenly split on what the best choice should be for potential designs. They were unsure which one they should go with and were discussing how best to pick. One group wanted the old tried-and-true method used in the past. Another group wanted to use a design based on a competitor's design that they thought was cool. Has this little scenario played out in your workplace in the past? I'm betting it has and that this sounds familiar to you.

I suggested using Personas to help with the decision. We pulled up the Persona of a casual job seeker the company had previously created. The casual job seeker Persona represented a large portion of the users of the company's systems.

The casual job seeker Persona was someone with limited domain expertise (or domain knowledge, meaning familiarity with the subject matter) who only occasionally needed to search for a job. Because this Persona had limited knowledge of the processes, terms, and procedures for searching for jobs, the design team felt creating a very simple set of basic interactions with lots of help icons and instruction text would be best. This design was based on the old tried-and-true method the company had been using.

And by referring back to that Persona both groups decided they had enough information to choose a design.

But wait!

We next pulled up another Persona representing the experienced job seeker. This Persona was based on people who were more experienced with looking and applying for jobs.

Although the actual number of users fitting this Persona was less than that for casual job seekers, the number of interactions this Persona typically had with the company's systems was much greater than that of the casual job seeker. Meaning, even though there were fewer experienced job seekers, they used these types of systems far more often than casual job seekers, which meant any system must include that Persona and their critical tasks.

The design team realized the basic model of interaction they had initially picked might not be the best choice. The advanced job seekers might find the tools too basic and might find the help text and icons useless or annoying. Rather than providing them with a seamless, efficient experience, the basic version might have been cumbersome and inefficient for this group. The design based on the competitor's interaction design, which was more advanced, seemed to be a better choice when considering this Persona.

The design team again faced a choice. One set of designs seemed to work better for casual users, the other set seemed to work better for experienced users. How to choose?

Using behavioral UX data, the team noticed that although experienced job seekers were fewer in number, the number of interactions they had with the system was far greater than the number of interactions with the casual job seekers. Clearly, the design team had to ensure experienced job seekers could have a satisfactory experience with the system.

In the end, using both of these Personas helped the design team to realize they actually needed two paths and interaction experiences. The experience was customized to enable advanced users to use the advanced tool, which had more features and functionality.

For this design, new users were provided with contextually triggered help information and a quick tutorial to show them how best to use the system. They were also provided with a way to choose a basic path, but with a large "Advanced" button that would easily bring the beginner user to the advanced tool if they so desired.

This new design and prototype was further usability tested with actual users who represented the Personas. Based on the results of testing, the system was further refined. Eventually the system had a single path and single tool but with the ability to customize the experience by hiding or unhiding help text, the tutorial flow, and advanced features.

The moral of this story?

A tough design decision was made easier by referring back to the Personas that would be using the system. The Personas helped the design team to focus on how typical users would expect to use the system. Using Design Personas clearly helped the team focus on making decisions that would empower and enable both sets of user groups.

In addition, adding in some quick usability testing with real users to validate the designs enabled optimization to happen during the development process.

This is true user-centered design at its best and a key reason why Personas are so helpful in the UX design process.

Aiding in Recruiting for Usability Testing

Another very important use for Personas is to help in identifying and recruiting people for usability testing. The Persona or Personas are critical for finding valid test participants for UX research projects.

The way it works it simple: you recruit people for usability testing who closely match your Persona. Testing people without the use of Personas means the usability testing data may be inaccurate. That's because the test might be with people who don't match the typical users and thus may not share the same domain expertise, critical tasks, mental model, and related attributes of your Personas.

So Personas are great for finding, recruiting, and testing people who will match the typical users of your website or app.

Easy peasy, right?

Personas and Remote Unmoderated Testing

There's not too much that could cause UX researchers to disagree, but one item that sometimes does just that is remote unmoderated usability testing. There are many fans of this tool, but there are a few who poo-poo it.

The common criticism of remote unmoderated usability testing among this group is that it uses "professional testers" who may not match the Personas of typical users of a website or app. This can be true, especially if the person creating the test doesn't take the time to use the recruiting and screening tools available to ensure that the testers match the Personas.

For those who are familiar with remote moderated testing, you can zoom ahead to the next paragraph. Perhaps just hang out and grab some virtual pizza; I'll be back with you in a minute.

For those of you unfamiliar with remote moderated usability testing, let's chat. And for both of you, I'll cover usability testing for UX optimization in more detail a bit later in this book.

Remote unmoderated usability testing is a type of UX research in which real people test critical tasks on a website remotely, such as from their house or anywhere they have Internet access and a mobile or desktop device. Typically the tester's voice and screen interaction are recorded. It's called unmoderated usability testing because the usability researcher is not present while the test is being conducted.

The way it works is testers are asked to perform certain tasks on the website or app while thinking out loud. Their screen interactions and voices are recorded while they conduct the test. After testing is complete, the researcher can view the resulting video. It's a very handy, fast, and scalable way to conduct usability testing.

Testing services like Loop11, TryMyUI, UserTesting, and Userlytics, among others, offer this type of remote unmoderated usability testing. Figure 4-2 shows an example.

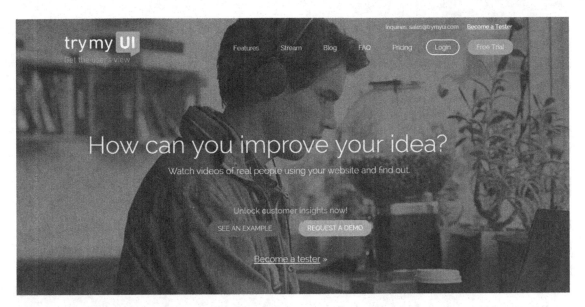

Figure 4-2. *TryMyUI is an example of a remote unmoderated usability testing service*

So why do some criticize remote unmoderated usability testing?

Because of the Personas, or to put it more precisely, the lack of ability to recruit testers who match Personas.

Some researchers feel the testers who are available with these services do not match the Personas of a typical website or app user. Some refer to these testers as "professional testers" because they believe the testers often test dozens or even hundreds of websites and apps.

In fact, most remote unmoderated usability testing services offer the ability to find, screen, and recruit testers who closely match Personas. If time and energy are put into screening and finding testers who match Personas, then remote unmoderated usability testing can be a very reliable source of UX research. It requires

1. Knowing the Persona or Personas to be tested

2. Creating a questionnaire or screener to identify testers who match the Persona

3. Recruiting and testing with only those testers

If the researcher takes the time to do the above, then finding testers who match the Personas for remote unmoderated usability testing is no more difficult than using any other recruiting method. And the Persona-matching testers who are doing the remote unmoderated usability tests will provide accurate results.

So not only are Personas good for aiding in recruiting for usability testing, they are also good for helping to settle arguments among UX researchers!

Personas and Scope Creep

Personas can help ensure teams are in alignment with which features and functions will be included in a website or app. This alignment can greatly help reduce scope creep or potentially negative changes in the design as the project is developed.

How?

By referring back to the users via Personas and using those Personas to help focus the team on who the website or app is being created for, what functions or features the Personas need, and how the Personas typically expect a process flow to work. These important decisions can be made by referring back to the Persona and asking

> *"Will this Persona be able to use this tool if we include or leave out X in this design?"*

And then there's the opposite of scope creep: minimum viable product (often called MVP) discussions. The goal with a minimum viable product is to create just enough of the functionality or features necessary for users to be able to achieve their goals, with little else in functionality available. Teams use MVP to get a product into market quickly, then to add functions and features to the product as time and resources allow.

MVP can be thought of as almost the reverse of scope creep. That's because it's the conscious decision-making going into removing functions and features to have a workable product that users find, well, useful.

Again, nothing beats having real users provide their input. But often there are so many smaller decisions that have to be made on a daily basis with these projects that having the design team refer to Personas when making decisions can be a great way to ensure the product will be useful to those it's designed to help.

Conclusion: Why Personas Matter

Personas are critical in UX and in any analysis of UX. Some of the more important reasons focus on helping to bring users and their needs to the forefront of any discussion of functions or features. Personas can

- Add context to UX behavioral data

- Enable user-centered design

- Aid in recruiting for usability testing

- Decrease scope creep

Personas are critical for determining what data to review as part of a UX behavioral analysis. Personas enable user-centered design by representing users when teams do not have the time or resources to bring real users in for testing, but still need to make decisions that impact the user experience.

It's very difficult to quickly and accurately recruit for usability testing without having a Persona. Using Personas and referring to them when making screeners to recruit testers helps ensure that the results of usability testing are more accurate because the testers match the most common users (Personas) of the system.

Personas can help teams reduce scope creep by being the litmus test for whether a feature of function is needed. And Personas can help when teams evaluate what a MVP experience should or should not include.

Now that you understand what a Persona is and why they matter so much, you're ready to have some fun! Let's go make a Persona, which in an amazing coincidence just happens to be the next chapter of the book! See you there.

CHAPTER 5

How to Create a Persona

The majority of you more than likely already have Personas that you can use to help with the data you'll be working with during the next chapters. Since you already have Personas, it's fine for you to skip this chapter and move right into the next chapter. Go grab your favorite beer, wine, coffee, tea or appropriate beverage for your tastes, and I'll see you there!

But for the rest of you, if you do not have Personas to work with, no worries! This chapter will help you create a Design Persona you can use as part of your UX analysis and optimization. So roll up your sleeves and let's have some fun!

Where Does Persona Data Come From?

The primary source of data for UX Design Personas is contextual inquiry. What is "contextual inquiry?"

> *Contextual inquiry is a user-centered design ethnographic research method in which the researcher finds the users on location and observes how they interact with systems in their own environment.*

I once worked with a product team that wanted to better understand how truckers searched for jobs. The team thought about the different types of scenarios and situations where they might be able to observe truckers in their context, such as while driving or taking a break at a truck stop. The product team went to a local truck stop with several questions in hand. They approached truckers and explained they were doing research to better understand how truckers think about jobs and how they look for a job.

So what do you think happened?

Do you think the truckers found the product team annoying?

Do you think the truckers avoided the team, not wanting to answer any questions from strangers?

© W. Craig Tomlin 2018
W. C. Tomlin, *UX Optimization*, https://doi.org/10.1007/978-1-4842-3867-7_5

Well, it turns out the truckers were thrilled that someone wanted their opinion, and they were more than happy to give the team their thoughts. They really enjoyed being able to tell the team all about how they go about looking for jobs. The product team came away with stacks of notes about the job search mental maps and habits of truckers. The information they gathered from this and other inquiries was used in the design of Personas and eventually early prototypes of the app that was developed for that market.

Without the contextual inquiry data gathered in the field, the Personas and resulting app might've taken a lot longer to get right. This is a perfect example of contextual inquiry and the process used to create a UX Design Persona.

So let's go through the steps it takes to put together a contextual inquiry for Persona development and what it takes to gather appropriate information during the inquiry. If you've never done contextual inquiry or field research before, this may seem a little daunting but, as with anything in life, all it takes is a little practice and you'll be an expert in no time.

How to Conduct Contextual Inquiry Research for UX Design Personas

So what are the steps necessary to conduct a successful contextual inquiry? How do you know what data to note when observing real users in their environment? And how do you consolidate your notes into cogent observations that design teams can use? It's just a matter of four simple steps.

Step 1: Prepare

Before heading out to go talk to real people, it's important to think about WHY you're going out into the field and WHAT you want to observe. Ask yourself several questions:

- Who should I be observing?

- Where should I be observing them?

- What do I want to observe?

- What questions do I want answered?

Write down your thoughts and questions onto a one handy sheet paper. You can refer back to it during your sessions.

It's all right to only have a few questions when you head out. Remember the primary mission is to observe. By doing so, you may find additional interesting information that you never would have thought to ask about.

But by preparing and having certain questions already thought through and written out in advance, you will find your contextual inquiry sessions will be more productive and more efficient in helping you find important data for creation of Design Personas.

Step 2: Get Out of the Office

Some people find this a little bit intimidating, but you actually have to leave your office and go where your users are. Sometimes this can be more difficult (say when trying to observe a doctor) but often it's easier than you may at first realize.

It can be difficult if your users are professionals who are typically not out in the public, such as doctors, lawyers, police, judges, and other professionals. For this kind of audience, it will be necessary to do some homework and reach out through recruiters to arrange appointments to visit these people in their offices or typical locations.

For websites or apps that are being designed for business people this can also prove to be a little bit difficult. I've found that your own network can sometimes be used to find these types of users. Most of us have friends and family who work at businesses other than our own. Using that network to help you find business people for your contextual inquiry is a good way to recruit people in those environments.

If you are building a website or app that is applicable to the general population, then you should have no trouble going out to where people are and recruiting them for your inquiry.

Coffee shops, outdoor or indoor malls, parks on busy weekends: you get the idea. There are plenty of places to find people who would be willing to spend a few minutes sharing their thoughts with you.

As to setting up the inquiry, sometimes it's as easy as just going to a nearby truck stop (if your users are truckers, that is) and asking people if they wouldn't mind answering a couple of questions. If your audience is more general, a great place to conduct contextual inquiry is a local coffee shop. Ask the person sitting near you for a few minutes of their time and offer them a coffee or a coffee gift card for their time.

A good rule of thumb for observing people in the context of their offices, homes, or other non-public locations is to arrange in advance to meet the participants. Scheduling sessions is efficient for their time as well as yours!

There are a variety of ways to invite people to participate in a contextual inquiry. What has worked well for me in the past is to avoid using any of the "stop" words when asking for a few moments of their time. What's a "stop" word?

They are words that may invoke negative emotions or responses from your prospective contextual inquiry candidates. Words like "test" or "study" or "observation" or even "contextual inquiry" may cause people to feel uncomfortable about what you are asking them to do. Nobody likes tests, and being "observed" may feel to some people like you are asking them to become human versions of laboratory rats.

Instead, ask for time to learn a "little bit more" about what that person does. Explain to them you're "hoping to get a better understanding of how they go about doing things" and "how they think about things." Explain that this information will be very helpful for the design you're doing but that no personal information will be captured or used. You can even offer to compensate them for their time with a gift card or other incentive, just as a way of saying "thanks for your time!"

This next part is very important!

I've found people are generally more than willing to share their thoughts with you **as long as they know you are not going to share their thoughts with a larger audience**.

Of course you will want to take all proper legal and privacy matters into consideration based on your place of employment and state/country laws regarding privacy. Measures should be taken to properly protect yourself, your firm, and the participant, including having your participant sign a non-disclosure agreement and a release to use their information for your UX research. Most firms, probably including yours, have these documents already and require that they be used, so it is often simply a matter of using these preexisting materials to make sure you are compliant with all of your firm's policies.

If you are prearranging to meet someone at their location, have a schedule in place to coordinate your contextual inquiries. I typically send out meeting invites to confirm the session, which is a great reminder for the participant that they've committed some time from their schedule for you. It's also a great way for them to reach out to you if something has come up and they need to reschedule the session.

Prior to the session, usually a day before, I like to send out a reminder (typically an email) reconfirming the session and asking the participant to contact me if they need to reschedule.

And just to be sure that I will obtain the proper number of contextual inquiries, I like to schedule one or two extra sessions, knowing that it's highly likely that one or two may end up cancelling.

How many sessions do I schedule?

It depends on the nature of the study I'm conducting. In general, I like to observe anywhere from 5 to 10 sessions to ensure I have plenty of opportunity to identify patterns of use or similarities in mental maps for process flows.

Step 3: The Session

So congratulations, all you preparation has worked well and you're now finally at your session! You remembered to bring your one sheet of questions, lots of paper to jot down your notes, and any nondisclosure or release agreements that your firm may require.

Here's where a little practice will go a long way. The better you get at being friendly, open, and interested in the participant, the better your sessions will go.

The best tip I can give you on conducting contextual inquiries is to practice listening, observing details, probing, and asking questions, especially "why" questions. This is because you want your participant thinking aloud, explaining how they go about conducting tasks, and what their mental map is for the process they go through.

To get started at the beginning of a session I like to ask a few warm-up questions. You can start by thanking the person for their time. Ask them questions about their background. If this is a work-related contextual inquiry, ask them about their job, why they got into it, and why they like it or don't like it. By using warm up questions you get to know a bit more about the person, plus you get them comfortable speaking with you.

If you are asking them to conduct a session on a website or an app, use the "show me" method. Ask them to show you how they do a task, and ask them to talk out loud as they go through the process of showing you.

During your session remember to follow these helpful guidelines:

- **Observe:** Just listen and take notes. Let your participant do most if not all of the talking.

- **Probe:** If someone says, "I like it because it's easier for me to use," ask "Why?" Drill down into the motivations and actions that ultimately cause satisfaction or dissatisfaction.

- **Follow Up**: If your participant makes nonverbal clues like raising an eyebrow, hesitating, scratching their head, or other signs of difficulty, be sure to follow up on those clues. "Gee, I noticed you seemed to hesitate there. Can you tell me what was going on?"

- **Ask Open-Ended Questions**: As a general rule of thumb, you want to only ask open-ended questions. These are questions that use words like "what," "why," and "how." A good one is "Can you tell me more about...?" Try to not ask close-ended questions like "Would you...," "Do you...," "Is this...?"

- **Take Copious Notes**: Make sure you write down lots of notes during your session. Although it may seem fresh while you're there, you'll find after your session that memory fades and important details may get left behind. Some researchers I know use digital audio or cameras via their cell phones to record the session. As long as you have the participant's permission to do so, and as long as your firm allows this, that's another helpful way to make sure you don't lose any important details.

- **Ending**: At the conclusion of the session, be sure to thank the participant for their time. Make sure you provide whatever incentive you've agreed to give them (gift card, free coffee, whatever). Ask them if it would be alright if you contact them to follow up later in case you need a bit of clarification on anything they shared.

Step 4: After the Session

Consolidate your notes immediately! Don't wait! Try to consolidate your notes as soon as possible after the session because what was fresh in your memory immediately after the session will start to fade with time. Your impressions and observations are key, so don't lose them by waiting too long.

Consolidate your notes by organizing them into an outline format with any potential important notes highlighted.

If you were part of a team that was doing contextual inquiries as a group, be sure to get together as a group as soon as you can to discuss your notes and observations, and review anything important that others may have noted.

It's usually best to have a single set of consolidated notes to represent the observations of the group if there are more observers in the contextual inquiry than just yourself.

Secondary Sources of UX Research Persona Data

There are other sources of data to help you pull together your Persona, but this is important:

> *These secondary sources do NOT replace primary source data (a.k.a. contextual inquiry data). They only add additional data points to that primary contextual inquiry data.*

Website or App Behavioral Data: If you already have a website or app for users and their critical tasks, you can learn a lot about existing behavior. A few questions this data can help answer include

- What content is or is not consumed?

- What click-paths are used to navigate?

- What search terms are entered in the search tool when people are looking for their solution?

Focus Groups and Surveys: Caution! Focus groups and surveys are far less reliable than other data collection methods. That's because what people SAY they do is very often different from what they ACTUALLY do. Still, if there is statistically significant data that is reliable, surveys or focus group data can sometimes be helpful as generalized inputs for the UX design persona. Focus groups are good tools for ideation, for generating ideas from users. Surveys are helpful tools if the data is statistically significant and relevant to the contextual inquiry. Just remember that focus groups and surveys cannot replace contextual inquiries; they can only add helpful data to them.

Customer Feedback and Voice of the Customer (VOC) Data: This data is also less reliable than contextual inquiry and actual behavioral data but still helpful for capturing user input. The issue with feedback and VOC data is typically input from the vocal minority is received, but not input from the silent majority. Also, this doesn't help you if you are building a new design or app that hasn't been used yet.

Marketing or Proto-Personas: Sometimes Marketing or Proto-Personas are already available. But remember, they are typically demographic or speculative (meaning based on secondary data) in nature. Although they may have some helpful information, you should not rely on them as a replacement for actual contextual inquiry. That said, review these Personas to see what information may fit or complement what you've observed from your contextual inquiries.

Next Steps After the Data Has Been Gathered

After you've finished gathering the contextual inquiry data, and any secondary data from the other sources, it is time to put it all together into a UX Design Persona.

Consolidate Data and Look for Commonalities and Patterns

Now that you have all your observations and the data discussed above, how do you synthesize it all into a Persona?

The key is to look for common patterns specifically in terms of how the user goes about accomplishing their goal or goals.

There are several steps to help you find patterns, which can include

- What does everyone say repeatedly about their goal or desire? What are they trying to accomplish?

- How does the current system help them accomplish their goal or goals?

- What parts of the current process work well?

- What parts of the current process do not work well?

- What consistent task-flow successes, or failures, are shared among the users?

- What are common pain points?

- What are consistent workarounds to existing problems?

- Is it common for people to have sticky notes with information on or near their computer to help them with their critical task? What's on those stickies?

- Do people frequently rely on cheat sheets or other non-system documents to complete a task? What are they?

- Where are there gaps in the existing process? What are those gaps?

By analyzing this information you'll start seeing patterns for goals and tasks, and for success and failure points in process and task flows.

Create Your Draft Persona (Hint: Work Backwards)

Now that you've completed your review of the various contextual inquiry sessions and you've identified your common patterns, it's time to start putting your Persona together.

Some practitioners start by rushing off to find a great picture and/or name for their Persona. However, I believe there's a better way to do this, and it's by working backwards.

Start with the end goal. What ultimately does the user want to accomplish with your product, website, or service? What's the desired end state?

Steps for Creating the Persona

The following are steps for creating your draft persona:

1. **Identify Critical Tasks**: What are the top 1-3 critical tasks necessary for the end user to be successful? This identification is also necessary for setting up usability testing. Sometimes the critical tasks for the user are based on existing systems, which may spark an idea to create a new way to help the user accomplish their goals. This is standard for entrepreneurs, who specifically look for new or better ways to help users accomplish their goals.

2. **Document Environment of Use**: Are there common places, devices, or third-party tools that are consistently used or needed to accomplish the above tasks? If so, and if they are important to the completion of the end user's desired end state, be sure to document them.

3. **Define Domain Expertise**: Is there a common domain expertise among the contextual inquiry participants you observed? Are there common familiarities or knowledge required with the systems, terminology, or processes used? If yes, document them. An example is a claims processor for a large insurance company who has to be trained on the terminology and processes before using an internal claims-entry system.

4. **Identify Pain Points**: What are the common pain points shared? What problems are consistent in their task flow?

5. **Create a Name**: Be sure to be culturally sensitive. Focus on common names that are easy to remember and that can easily be used by your team. Names are important. Don't scrimp on spending time to find just the right name for your Persona!

6. **Find a Picture**: As humans, we are visual creatures, so a face and name are important to humanize the Persona. I recommend real pictures vs. cartoons or clip art, and if possible use pictures showing the end user in context of use of the system. For example, if you're creating an app to find a lost dog or cat, a picture of a happy pet owner hugging their pet would be a good choice.

Common Attributes of UX Design Personas

Like snowflakes, no two formats for Design Personas are exactly the same across firms who use Personas. But when you've finished your draft Design Persona, you'll probably see several common attributes that are shared among most Design Personas, no matter who has created them.

There are thousands and thousands of variations of Personas out there; just do a search for "UX Persona" in Google images to see what I mean (Figure 5-1).

Figure 5-1. *The very wide variety of different types of Personas found in a Google image search for "Persona"*

However, for UX design, research, and usability testing purposes, most Personas should share the same basic attributes in common (including yours):

- **Picture**: Pictures are important because they personalize and humanize your Persona. They help tell the Persona's story, so they MUST be an accurate visual representation of the Persona. Don't just use any random picture. Spend time on finding and using the right picture to humanize your Persona.

- **Critical Tasks**: Typically no more than three.

- **Scenario**: Specific to the critical task or tasks, what is this Persona trying to accomplish?

- **Background**: The background for the scenario. Why is this Persona trying to accomplish a critical task?

- **Devices**: What device or devices does the Persona typically use, or what third-party tools are required? This is less important now because most people can and do use multiple devices; however, it's still important especially if you are discussing specific software or other solutions that require specific devices.

- **Domain Expertise**: How educated or familiar is the Persona with the subject matter, terminology, and existing process flow? Do they have a good understanding of terminology and a solid mental map of how the process and task-flow should work, or not?

- **Environment**: Where are your users when they are conducting their tasks? This is somewhat less meaningful now that the Internet is everywhere via mobile devices and tablets in coffee houses, at work, etc. Yet it's still important to consider the environment in which the user is engaged with your website or app.

If you search for Personas on the Internet, you will see a huge variety in styles and types of Personas, from the very detailed to the very basic. But for UX design and research purposes, you can't go wrong by making sure you have the above data clearly defined for your Personas. Figure 5-2 highlights some of the common elements most personas share.

Persona #2 - Joel

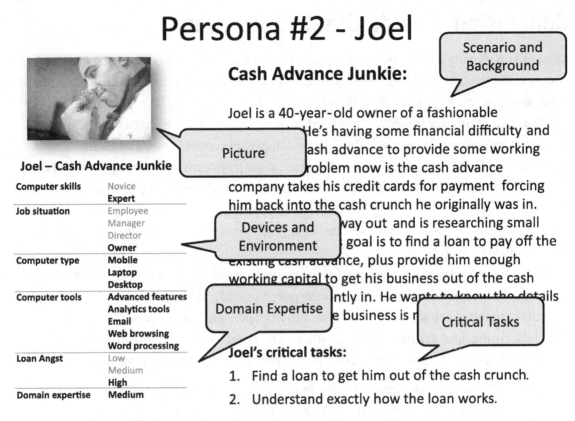

Cash Advance Junkie:

Scenario and Background

Joel is a 40-year-old owner of a fashionable ~~shop~~. He's having some financial difficulty and ~~took a~~ cash advance to provide some working ~~capital. His p~~roblem now is the cash advance company takes his credit cards for payment forcing him back into the cash crunch he originally was in.

Picture

Devices and Environment

~~He's looking for a~~ way out and is researching small ~~loans online. His~~ goal is to find a loan to pay off the existing cash advance, plus provide him enough working capital to get his business out of the cash ~~crunch it is curre~~ntly in. He wants to know the details ~~of how the onlin~~e business is run.

Domain Expertise

Critical Tasks

Joel – Cash Advance Junkie

Computer skills	Novice
	Expert
Job situation	Employee
	Manager
	Director
	Owner
Computer type	Mobile
	Laptop
	Desktop
Computer tools	**Advanced features**
	Analytics tools
	Email
	Web browsing
	Word processing
Loan Angst	Low
	Medium
	High
Domain expertise	**Medium**

Joel's critical tasks:

1. Find a loan to get him out of the cash crunch.
2. Understand exactly how the loan works.

Figure 5-2. *Common elements most Personas share*

If all this seems difficult, it is and it isn't. There IS a lot that goes into creating a Persona. Some people have great careers spending their working days doing nothing but Persona research and development work.

My suggestion is you try this several times by doing practice sessions. You can use your family or friends. Try out your technique of observing, asking follow-up questions, and noting patterns. It doesn't have to be about your particular website or app; any critical task could be used for your practice session.

The reality is Persona research is really just about observing people in their environment, asking questions, and learning more about them and the tasks they are trying to accomplish. And that is one of the beautiful things about Persona work: it brings you that much closer to the people you are designing for.

Conclusion: How to Create a Persona

Most of you already have Personas that are ready to use. But for those of you who don't, you now know enough to be able to go out, observe people in their environment, and use your observations to put a Design Persona together.

Contextual inquiry is the method used to observe actual users in their environment to learn how they go about tasks, their mental map for process flows, and to ask questions and learn more about how they use existing systems. It means getting out of the office and going to the places where your users are and observing them as they use the systems you are researching. It's also a good idea to review secondary sources of information about users, always remembering they are secondary and cannot replace the contextual inquiry data.

The steps necessary for a successful contextual inquiry include doing the preliminary work of asking yourself questions about the purpose and goal of your sessions. During the contextual inquiry sessions, you'll want to ask open-ended questions, listen more than talk, probe, and ask follow-up questions. You need to take copious notes!

After your session, you should consolidate your notes and begin looking for patterns.

- What do the participants share in common about their goals or desires? What are they trying to accomplish?

- How does the current system help them accomplish their goal or goals?

- What parts of the current process work well?

- What parts of the current process do not work well?

- What consistent task-flow successes, or failures, are shared among the users?

- What are common pain points?

- What are consistent workarounds to existing problems?

- Is it common for people to have sticky notes with information on or near their computer to help them with their critical task? What's on those stickies?

- Do people frequently rely on cheat sheets or other non-system documents to complete a task? What are they?

- Where are there gaps in the existing process? What are those gaps?

As you create your persona you'll want to include several elements of important information including:

1. **Identify Critical Tasks**: What are the top 1-3 critical tasks necessary for the end user to be successful?

2. **Document Environment of Use**: Are there common places, devices, or third-party tools that are consistently used or needed?

3. **Define Domain Expertise**: Is there a common domain expertise, meaning familiarity with the systems, terminology, or processes?

4. **Identify Pain Points**: What are the common pain points shared among users?

5. **Create a Name**: Be sure to be culturally sensitive. Focus on common names that are easy to remember and that can easily be used by your team.

6. **Find a Picture**: As humans, we are visual creatures, so a face and name are important to humanize the Persona.

Creating Personas takes work. If you've not done it before, you may want to practice a few times with family and friends.

Once you've tried a few sessions, I think you'll agree with me that it is very rewarding to spend time with your users, learning more about them, and that the time you spend will in turn make your design and development efforts that much better.

Now that you've learned about Personas, why they are important, and how to create them, it's time to move to the next step: applying Personas in your UX research. The primary point in having Personas is to use them to help you narrow down what data you need to evaluate to help you identify ways to improve the user experience of websites. And I'll cover that in the next chapter!

CHAPTER 6

Behavioral UX Data

Sometimes too many choices can be a bad thing.

Recently I was instructed by my wife to go to the grocery store and buy laundry detergent and a few other things. Now you would think this would be a simple task for a guy with a university degree, reasonable intellect, and deep domain experience at finding things, right?

Wrong.

When I got to the store and found the laundry detergent aisle, I was dismayed to see row after row after row of laundry detergent. It was bewildering. There were dozens and dozens of variations of detergent. Worse, there were subtle differences in each variation of detergent. Which one was the one I was supposed to buy?

Among the many, many choices I had there was Fresh Scent, Irish Fresh Scent (causing me to ponder if the Irish have a different fresh scent than the rest of our fresh scents), Clean Breeze, Lemon Scent, No Scent, Ultra Stain, Free and Gentle, Cold Water Clean, Oxi, Simple Clean, Detergent with Bleach, Detergent without Bleach, HE Formula, Regular Formula, Pods, Liquid, Concentrated, etc., etc., etc. Figure 6-1 is a picture I took clearly presenting the evidence for why I was so thoroughly and completely confused.

© W. Craig Tomlin 2018
W. C. Tomlin, *UX Optimization*, https://doi.org/10.1007/978-1-4842-3867-7_6

Figure 6-1. _Sometimes too many choices can be a bad thing_

I was beyond bewildered and confused. With so many choices and so much data on all the variations I had available to me I was completely overwhelmed.

So what was my decision?

You may have guessed it.

I made no decision.

I did not buy the detergent. Instead, I purchased the other items I had been requested to buy and went back home without the all-important laundry detergent.

Has that ever happened to you?

More than likely, it has. And there's a reason for that.

In his Ted Talk and book, _The Paradox of Choice: Why More Is Less_, psychology professor Barry Schwartz provides insights into why choice, so highly valued in our culture, is actually causing us to be unhappy. Rather than providing us the freedom and happiness we seek, too many choices are sometimes causing us to feel we need to make decisions that in the end make us feel worse.

There are a variety of studies that have demonstrated that offering too many choices sometimes causes people to make no choice. It sounds odd, but limiting choices can actually empower people to make a choice.

All of which causes me to say...

Behavioral UX data is like laundry detergent. There are so many choices available that deciding what to use can be overwhelming.

So with that cautionary tale in mind, let's dive into the world of behavioral UX data.

Behavioral UX Data Overview

Behavioral UX data is the "what's happening" data. Because this data is quantitative in nature, it is used to identify what behaviors are, or are not, occurring on your website. This information is critical for any analysis of website or app optimization opportunities.

We use behavioral UX data to determine what types and amounts of interactions are happening on the site or app.

This quantitative data, or WHAT data, coupled with the WHY data coming from UX and usability testing gives us a comprehensive view, or what I refer to as a 360-degree view, into website activity.

When analyzed together, the behavioral UX and usability testing data provide the 360-degree view into what's happening and why it's happening on the website. This more educated and enlightened view into website engagement makes for far more informed decisions as to where website issues are, why they are happening, and what to do about them. This leads to better optimization recommendations and improved website conversion.

Behavioral UX data is critical, but by itself it's not enough to make informed analysis and optimization recommendations. Think about it this way:

Behavioral UX data is the sign-post for WHAT is happening on a website, but not WHY it's happening.

Sources of Behavioral UX Data

So where does behavioral UX data come from? There are several common sources for this data including

- Website Log Analytics Programs (Google Analytics being the most common)

- Advertising Systems (Google Adwords, Facebook Ads Manager, Hootsuite, etc.)

- Content Management Systems (often abbreviated as CMS, examples being WordPress, Drupal, Magento, etc.)

- Marketing Automation Systems (Eloqua, Marketo, Pardot, etc.)

- eCommerce Systems (BigCommerce, Shopify, Volusion, etc.)

- Custom Back-End Systems (designed to complete tasks, purchases, transactions)

It can be overwhelming to consider all the possible sources of behavioral UX data. And if you consider all the data available in each of those systems, it can be even more overwhelming! The important thing to remember is

The goal of using quantitative data from these systems is to help answer specific questions about what's happening in the user experience.

What sort of questions you may have and how you use this data is the subject you will explore later in the book. For now, the key point is that once you know your questions, and once you've determined what sort of data you need to answer those questions, it's highly likely that one or several of these data sources will be the place you can go to get that information.

Starting with the questions first, and having a sense of the data you need to answer those questions, will reduce that overwhelming feeling you may have due to all that data. And that will make it easier for you to concentrate on the behavioral UX data analysis.

Now that you have a sense of the sources, let's look at each of the various types of behavioral UX data available and how they can be used to answer your questions.

Types of Behavioral UX Data

There are four broad types of behavioral UX data:

1. **Acquisition** (PPC keyword data, etc.)

2. **Conversion** (actions such as clicks, sign-ups, downloads, etc.)

3. **Engagement** (such as bounce rate, time on page, etc.)

4. **Technical** (visits by browser, screen resolution, etc.)

Let's review each type in more detail.

Because Google Analytics is a very popular and widely used analytics platform I'll use the reports from it for the brief overviews below. But other web analytics programs like IBM Digital Analytics or Adobe Analytics will offer similar types of data and reports, although they may be referred to by different names.

Acquisition Data

Acquisition data is useful for determining where people came from when they decided to visit your website or download your app. Knowing where people came from, what they were looking for, and whether they found it (or not) is a critical element of website optimization. The following sections cover several examples of acquisition data and how they can help with optimization.

Organic Search Keyword Data

Organic search keyword data means any keywords people entered and clicked on when they came to your site from the non-paid portion of search results pages. Google and other web analytics tools will provide this information typically in a SEO (Search Engine Optimization) report.

Organic search terms are very important for identifying what keywords or phrases people are searching for on search engines when they found your site and clicked to visit. They are also important because they often represent the majority of traffic to a site that converts (i.e., completes an action on the site).

Figure 6-2 is an example of a typical organic search report showing the keywords entered (search query), the number of clicks, and the number of impressions on the search engine results page the terms generated.

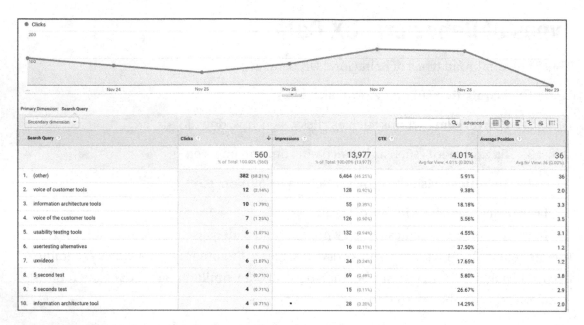

Figure 6-2. *Organic search report available in Google Analytics*

Paid Media Data

Paid media means forms of advertising that are not paid search, such as video ads, display ads, social media ads, and (the still present but almost completely ignored) banner ads.

Knowing which paid media ads and the content in those ads your viewers engaged with to visit your site is very helpful contextual "what's happening" data. Much like the organic search data, paid media data can shed light into what concepts, content, or text is triggering a response in your viewers. This data can help you identify what caused visitors to come to your site.

Other quantitative data points can be combined with paid media data to determine whether those visitors who clicked a paid media ad stayed on the resulting website page or if they immediately bounced away from that page. Knowing this helps to shed light into whether the paid media ad is properly setting expectations for what content visitors will eventually find, should they click the paid media ad.

Paid Search Advertising Keyword Data

Paid search keyword data refers to pay-per-click ads (known as PPC) available on search engines such as Google and Bing. PPC ads are the ads on search engine sites that display advertisements in the upper and lower areas of the results pages for keyword searches. Other types of paid search ads include text link ads and related types of paid search-based advertising on sites other than the main search engine sites.

Paid search advertising keyword data is very helpful for identifying what terms people were searching for or what they clicked when they came to your site. Among other things, it tells you what specific terms or phrases were being searched for, how many of those terms were entered (impressions), and how many people clicked those terms to visit your site.

Figure 6-3 demonstrates a typical PPC keyword report with data. Knowing what terms people are searching for and what they are clicking to visit your site are good sources of "what's happening" data.

Search keyword ⇌	Campaign ⇌	Ad group ⇌	**Impressions** ⇌	Clicks ⇌	CTR ⇌
usability test	Usability Test 2	Usability Testing Phrase	1,979	3	0.15%
usability	Usability Test	Usability Test	1,356	2	0.15%
Usability testing	Usability Test 2	Usability Testing Exact	1,052	5	0.48%
user interface	Usability Test	Usability Test	711	0	0.00%
human factors	Usability Test	Usability Test	604	0	0.00%
jobs usability	Usability Test	Usability Test	590	1	0.17%
heuristics	Usability Test	Usability Test	494	0	0.00%
usable	Usability Test	Usability Test	489	1	0.20%
software engineering	Usability Test	Usability Test	388	1	0.26%
user interfaces	Usability Test	Usability Test	357	0	0.00%
information architect	Usability Test	Usability Test	321	0	0.00%

Figure 6-3. PPC Keyword Report from Google Adwords

Referral Data

A referral is a visit to your website from any referrer source. Referral means "where did your website visitors come from?" Did they come from a search engine through a search results page linking to your site? Perhaps they came from a social media link to your site. Maybe they came from a related topic website.

Knowing the referral data about where people came from when they visited your website is another helpful source of "what's happening" data because it provides context to where people were when they came to your site. This is useful information for understanding the context of their visit to your site. Were most visitors coming from a search engine? Were they coming from a competitor website? Perhaps they were coming from other pages in your site (such as going from an internal page to the home page). Understanding how many visits came from each referral source can help shed light into where visitor came from, where they went on your site, and potentially whether they found what they were looking for or bounced away.

Figure 6-4 displays a report from Google Analytics detailing the various social media sources of traffic to the site and the data associated with that traffic.

Social Network	Sessions ↓	Pageviews	Avg. Session Duration	Pages / Session
1. Twitter	1,742 (41.22%)	2,791 (43.91%)	00:01:03	1.60
2. LinkedIn	835 (19.76%)	975 (15.34%)	00:00:35	1.17
3. reddit	729 (17.25%)	1,406 (22.12%)	00:00:52	1.93
4. Facebook	577 (13.65%)	681 (10.71%)	00:00:50	1.18
5. Blogger	105 (2.48%)	157 (2.47%)	00:00:56	1.50
6. StumbleUpon	68 (1.61%)	69 (1.09%)	00:00:06	1.01
7. Stack Exchange	67 (1.59%)	79 (1.24%)	00:00:42	1.18
8. Pocket	25 (0.59%)	36 (0.57%)	00:00:48	1.44
9. Pinterest	23 (0.54%)	86 (1.35%)	00:06:06	3.74
10. Ning	17 (0.40%)	21 (0.33%)	00:00:22	1.24

Figure 6-4. *Social Network Report from Google Analytics with social traffic sources and data*

Source/Medium

The Source/Medium Report in Google Analytics is a helpful way to identify which acquisition source and medium are sending traffic to your website. This is useful for identifying how much each of your acquisition channels are contributing toward total traffic to the site and which medium in those sources is providing the visitors.

Think of source as the type of channel (i.e., Google, Yahoo, Bing, etc.) and the medium as the type of traffic that source is sending (i.e., organic traffic, referral traffic, paid traffic, etc.). Figure 6-5 shows a Source/Medium Report with the details for each source.

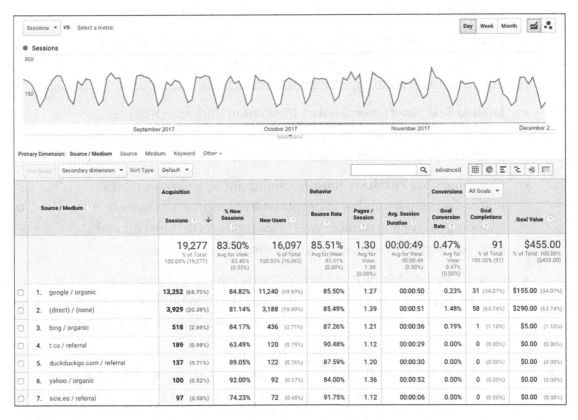

Figure 6-5. *This Source/Medium report displays the top sources and mediums generating traffic to the site*

Conversion Data

Conversion data is an important behavioral UX data set and is typically what your firm cares about a LOT. Conversion data is used by Marketing, Product, Sales, and Support teams to evaluate how well the website or app is attracting and "converting" visitors into taking various types of actions.

There are many possible types of conversions. Here are just a few:

- **Click-Through Rate (CTR)**: The percentage of people who see an ad or link and click it

- **Download Rate**: The percentage of people who download an app or other file

- **Lead Conversion Rate**: A catch-all term for any type of conversion that occurs once a lead takes further action on the site such as downloading a white paper or completing a form

- **Lead to Sale Rate**: Often used on Business to Business (B2B) websites where the percentage of the number of leads that end up purchasing a product or service is measured

- **Shopping Cart Purchase Rate**: An important eCommerce metric that measures the number of people who place products in their shopping cart vs. the number who place products in their cart and then actually check out and purchase them

- **Suspect-to-Lead Rate**: A marketing metric referring to visitors to your website who are initially unidentified but then provide some identification information on a form to become a known "lead"

Figure 6-6 is an example of a Conversion Report that highlights three separate types of conversions happening over time. This type of reporting is often used by marketing and product teams to measure the activity occurring on a site for specific periods of time.

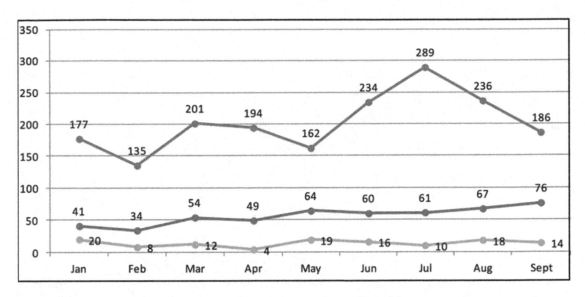

Figure 6-6. *Conversion Report showing number of leads generated per month with three sources trended over time*

There are many other forms of conversion data. Most are more specific to the particular needs of the firm in measuring what activity is occurring as a user takes various actions on their website or with their app.

Engagement Data

The following sections cover examples of engagement data.

All Pages

All Pages is a ranked listing of the pages with the most to least pageviews for a specific time period. This information can be helpful for identifying pages that users are engaging with (i.e., visiting) and those they are not engaging with.

Trending the ranked positioning of a page before and after content or navigation changes can help identify if the optimizations are working, by sending more traffic to the page or not. Figure 6-7 demonstrates a typical Pageviews Report based on the top pageviews for each page, ranked in order from higher to lower for the given time period.

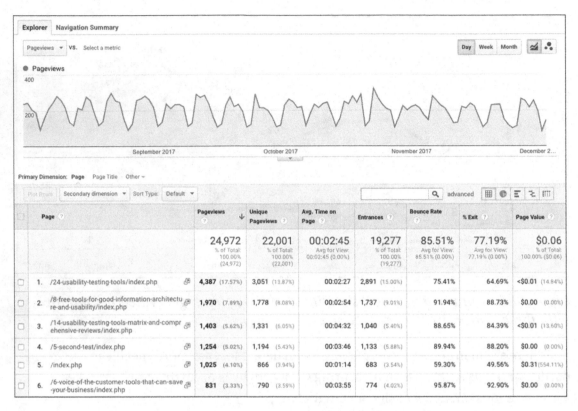

Figure 6-7. *A Pageviews Report in Google Analytics lists in ranked order the top pages by pageviews*

Behavior Flow

The Google Analytics Behavior Flow Report visually presents the top paths users took from one page to the next. This report is very helpful for visually identifying the most common flow or paths your visitors are taking when engaging with your site. It also helps identify what pages they most commonly visit on their journey in your site. Figure 6-8 displays a typical Behavior Flow Report.

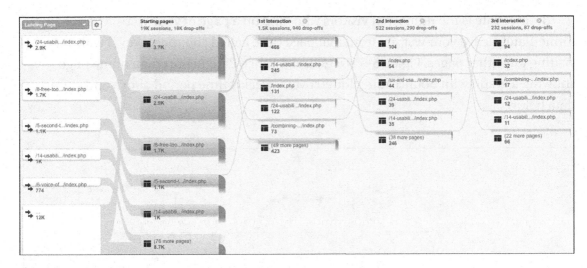

Figure 6-8. *The behavior flow in a typical website demonstrates the path visitors took*

A Behavior Flow Report is a useful tool for evaluating whether people are taking the desired path to content and by how much. In a perfect world, the top paths you would like your visitors to take would be the most visited pages, thus in this report it would be the very top row of pages going from the initial landing page on the left all the way through to the second, third, or potentially fourth interaction on the right. Seeing whether this desired path is or is not displayed at the top in this report is a quick way to determine if your navigation, information architecture, and labeling are working as expected.

If a page you are not expecting is ranked in the top row (for example, a search results page in your website), then this may be a bad thing, indicating visitors are not finding what they are looking for and therefore are searching for it. This often requires further follow-up using qualitative data to determine why your visitors are not finding the information they are looking for.

Bounce Rate

The bounce rate for content pages can tell you how well your content meets the expectations and needs of your website visitors.

Bounce rate is the number of visitors who land on a page of your website and then immediately leave your website from that page (also called "bounce away") without visiting any other pages. Figure 6-9 is a typical Bounce Rate Report for a home page showing a graph with bounce rate plotted over a weekly time period above and summary data below.

Figure 6-9. *Bounce rate for a home page with a graph at the top and summary data below*

This website behavior often occurs if website visitors thought they were going to find certain content when they came to a page on your site, but when they got to that page they did not find what they were looking for, and so they left your site. In other words, they bounced away. Bounce rate can also help you determine whether your website navigation and labeling are effective, or ineffective, for directing website visitors to the correct page of content.

Click Heatmaps

A click heatmap is a visual representation that defines where people are clicking on a page. The heatmap is an overlay that uses hotter (red, yellow) colors to represent higher click areas and cooler colors (blue or no color) to represent lower or no click areas. Figure 6-10 is an example of a heatmap of a home page.

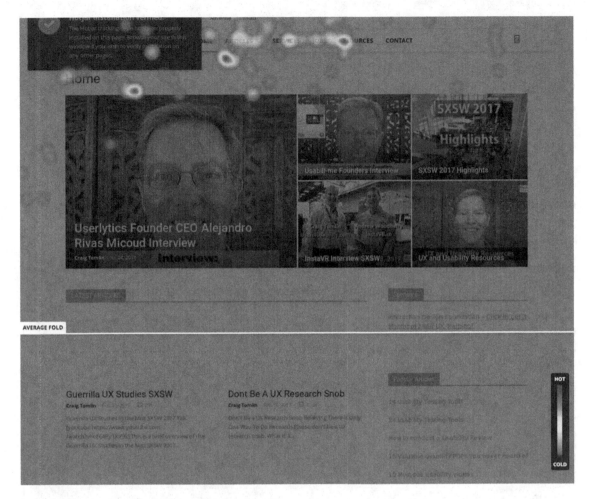

Figure 6-10. *A heatmap of a homepage*

There are a variety of tools that produce visual heatmaps including ClickTale, CrazyEgg, and Hotjar. Using one of these tools enables you to determine whether the objects you want visitors to click are working or not. It also shows the objects visitors are clicking that they should NOT be clicking.

I've found that graphic design treatments that look like clickable objects, but are not, often attract undesired user clicks, which is counterproductive to the user experience. Likewise, if clickable objects are designed such that they do not appear to be clickable, this can cause poorer click activity.

The principle that determines whether a graphic image looks clickable or not is known as "visual affordance." This means, does the object have a visual appearance that helps a viewer understand the object can be clicked. An example of visual affordance is the blue underlined text on a web page. The blue underline is a visual affordance that we have learned means the text is most likely a hypertext link.

Using heatmap data to understand how well visual affordance is or is not working for clickable objects is a great way to hunt down those problematic areas of a design that are causing poor interaction.

Exit Pages and Exit Rate

Exit pages are the last page a website visitor is on before leaving the website. The definition in Google Analytics is the count of the number of pageviews that exited the site from that page.

For all the pageviews on a certain web page, the exit rate is the percentage of the pageviews on that page that were the last page on the website. Figure 6-11 shows the Google Analytics Exit Pages Report that provides both number of page exits and the exit rate.

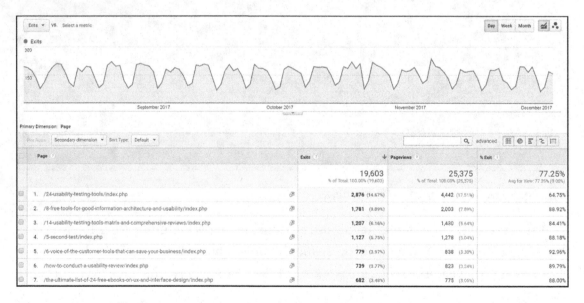

Figure 6-11.   *The Exit Pages Report provides the listing of the last page Google identified before visitors left the website*

Exit rate is different than bounce rate. Bounce rate means only that one particular page was visited before the visitor left the website. With exit page and exit page rate, this implies that other pages were visited on the site prior to the exit from the page in question.

High numbers of exits from a page may not necessarily be bad. It's good if that page is the end of a task flow in which the user accomplished their task. It's bad if the exit is in the middle of a task flow, implying users are not accomplishing their tasks.

For example, a high exit rate on the payment confirmation page of an eCommerce website is potentially really good. A high exit rate in the beginning of the checkout process on an eCommerce site is potentially really bad.

Landing Pages

The landing page is a page-specific metric that is useful for counting the number of times visitors first visit a certain page on your website. This is helpful for evaluating whether users are coming to a specific page on your site.

This data can be trended over time to evaluate how well audience acquisition campaigns are working in engaging your audience and bringing them to a unique page or pages on your site. Figure 6-12 provides information on the top landing pages as well as additional engagement information regarding that page.

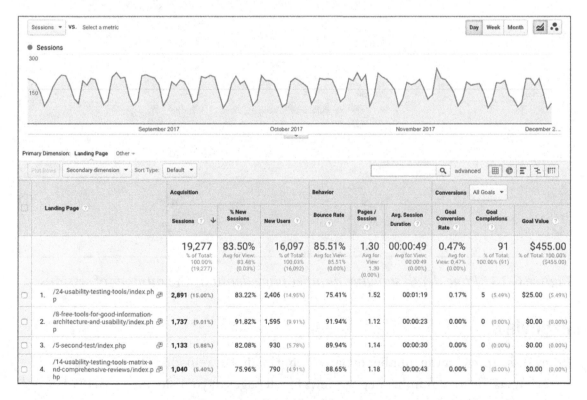

Figure 6-12. *The Top Landing Pages Report identifies the number of sessions per page and related engagement metrics for that page*

Page Depth

Page depth is a high-level metric that is an average of the total number of pages visited per session for a given time period. In theory, a higher number of pages visited per session is associated with a better user experience and engagement with the content.

But in fact, this greatly depends on the website and the amount and type of content available. Sometimes it's a good thing for people to only visit one page in a session. For example, if you are sending paid media traffic to a landing page, it would be fine if there was only one page visited.

Page session duration and depth will also vary depending on the goals the users have when visiting your site. If you have a publication site with many articles on a subject, you might hope and expect to see more pages visited per session because in theory visitors are finding related content they can read about a topic.

Figure 6-13 shows a chart displaying the page depth by the number of user sessions.

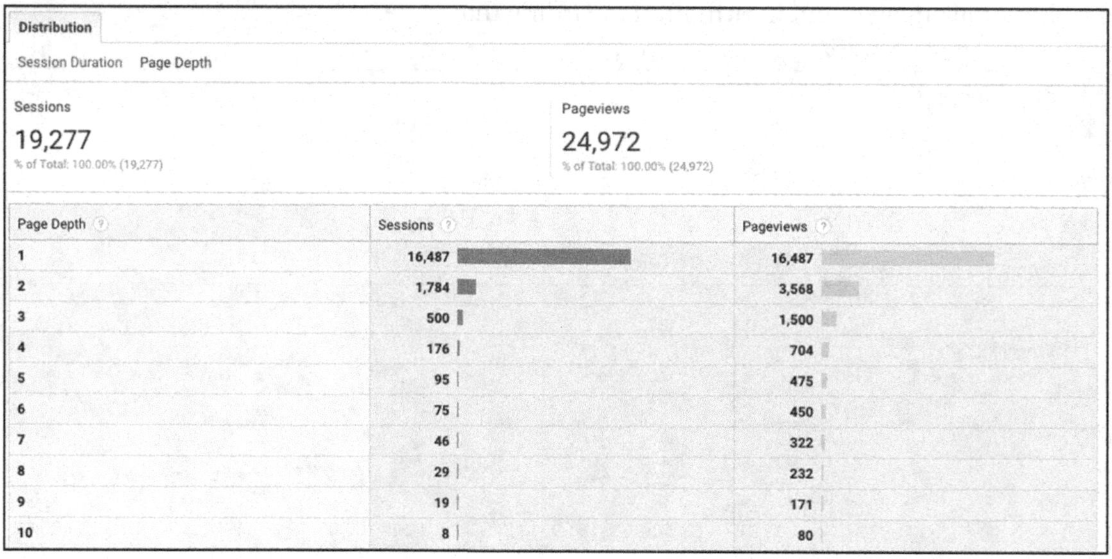

Distribution		
Session Duration Page Depth		

Sessions	Pageviews
19,277	24,972
% of Total: 100.00% (19,277)	% of Total: 100.00% (24,972)

Page Depth ?	Sessions ?	Pageviews ?
1	16,487	16,487
2	1,784	3,568
3	500	1,500
4	176	704
5	95	475
6	75	450
7	46	322
8	29	232
9	19	171
10	8	80

Figure 6-13. *Page depth showing number page depth by the number of sessions*

Pageviews

Pageviews represent the number of pages being loaded or reloaded in a browser. Pageviews is the total number of pageviews for all pages in your site for a specified amount of time (such as over a day, week, month, year, etc.).

This metric is a high-level metric focused at the total page activity for your site. Remember that a single user can have more than one view of a page during their visit to the site, so higher numbers of pageviews do not necessarily mean higher numbers of users to the page. Figure 6-14 shows the pageviews for a website on a daily basis. Note the dips in pageviews caused by the weekends.

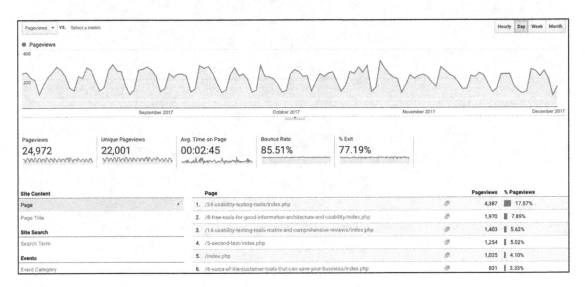

Figure 6-14. *A Pageviews Report shows the total number of pageviews per day*

Scroll Heatmaps

Like click heatmaps, scroll heatmaps are a visual overlay of data representing how far down a screen your users scroll. Areas on a page that are viewed more often are displayed with a hot color like red or yellow. Areas on a page that are not viewed often, or at all, are displayed with a cooler color like green, blue, or no color. Tools like ClickTale, CrazyEgg, and Hotjar offer scroll heatmaps.

Figure 6-15 is an example of a home page with the scrolling heatmap overlay. Note the location of the average page fold, and the decreasing number of users who do not scroll further down the page. Near the top of the page there are red and yellow colors, indicting a high percentage of measured visitors viewed the page. Further down, the colors shade to yellow, green, and then blue, indicating that fewer visitors viewed those sections of the page.

Figure 6-15. *Scroll heatmaps visually present screen resolution and scroll data*

In the past, there were some ~~heated arguments~~ discussions in the UX field about whether the fold (the area of the page below the browser window) was or was not important. Data from a university study seemed to imply the page fold was actually not important or real because the people who participated in the study actually scrolled well down the page to read the content. To which I say, check out your scroll heatmaps! You may be very surprised at the lack of scrolling people actually do on a page.

Note that there can be big differences between scrolling on desktop versions of websites vs. mobile versions. Either way, having this scroll heatmap data is a great way to evaluate how much users are scrolling down on your pages and what content they are or are not viewing.

Sessions

A session is a unique set of interactions a single user takes in a given amount of time on your website. Google Analytics defaults that time frame to 30 minutes. So whatever the user does on your site, such as visiting pages, downloading files, viewing videos, all their interactions in that 30 minute time period are considered a single session.

Sessions are a higher-level metric that can help you determine whether activity on your website is going up, down, or staying the same over time.

Figure 6-16 is an example of session report in Google Analytics with a line graph of sessions per day (note the pattern of dips in sessions during weekends, which is very common for almost all websites).

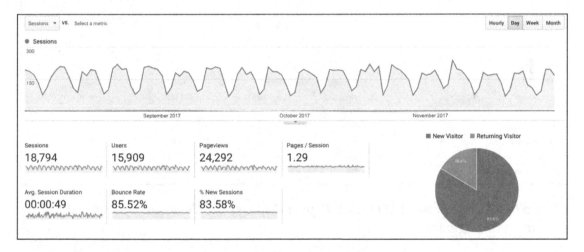

Figure 6-16. *Sessions Report from Google Analytics*

Session Duration

Another high-level metric is average session duration, which is defined as the total duration of all sessions (in seconds) divided by the total number of sessions. This can be a handy way to evaluate how well the content of a website may be engaging the user.

Identifying if there are many sessions or pageviews falling within a very brief session duration time period may help identify if visitors are not finding the information they seek or are abandoning the site without exploring additional pages. Trended over time, this data can help identify if content and/or navigation optimizations are having the intended effect of improving the engagement with the site.

A note of caution: In Google Analytics, Google cannot tell how long a visitor stayed on the final page they visited before leaving. There is a technical reason for this that I won't get into in this book. This means session duration could be underreported, sometimes vastly underreported.

For trending engagement over time for the whole site, this is a handy metric, but for unique pages, see the time on page metric, which will probably be more useful.

Figure 6-17 provides an example of a Session Duration Report. Note that session duration is NOT the same as time on page, which evaluates how long a user visits a unique page on the site.

Session Duration	Page Depth			
Sessions			**Pageviews**	
19,277			24,972	
% of Total: 100.00% (19,277)			% of Total: 100.00% (24,972)	

Session Duration ⑦	Sessions ⑦		Pageviews ⑦	
0-10 seconds	16,823	▇▇▇▇▇▇	17,179	▇▇▇▇▇▇
11-30 seconds	421	▏	1,005	▏
31-60 seconds	359	▏	988	▏
61-180 seconds	583	▏	1,964	▎
181-600 seconds	579	▏	1,849	▎
601-1800 seconds	457	▏	1,514	▎
1801+ seconds	55	▏	473	▏

Figure 6-17. *The Session Duration Report displays the duration by number of sessions and pageviews*

Time on Page

The time on page metric is an average of the total time on that page divided by the total number of pageviews of that page. As with session duration, Google cannot tell how long someone stayed on the last page of their visit to your site. So as long as this page is not the last page they visited (i.e., has a low exit rate) then this metric will be fairly accurate.

Figure 6-18 shows a typical Time on Page Report with the averaged time on page for all pages in the graph plus unique page data in the table below.

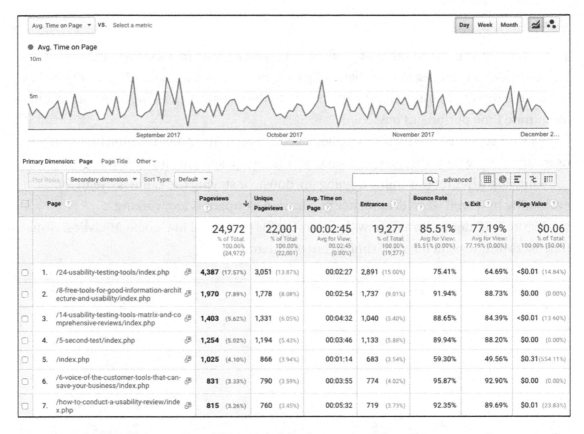

Figure 6-18. *A Time on Page Report with the average for all pages at the top and unique page data in the table below*

Time on page is one of the more useful behavioral UX metrics for evaluating the engagement on a website. Assuming you know how long it takes for the average reader to consume content on the page, you can determine whether your visitors are reading the content, or not, based on the average time on page.

A note about time on page: Do not assume pages with higher time on page metrics means those are "better" pages. Remember that pages with little content will produce lower time on page metrics vs. pages with greater amounts of content. So, lower time on page numbers for those pages is actually not a bad thing.

Likewise, pages that should have lower time on page numbers but are actually experiencing higher time on page numbers than expected might mean your visitors are having trouble finding what they are searching for.

Users

The users metric is the number of users visiting your website in a specific time period. This is different than sessions in that a single user is only one user, but that user may have more than one session if they come back to the website sometime after the default 30-minute time period has passed.

As with sessions, this metric is higher level and helpful for determining how many users are visiting your site, or a page on your site, over a specific time period.

Knowing whether this traffic is going up, down, or staying the same is useful for evaluating how well website audience acquisition campaigns are working.

Figure 6-19 demonstrates how the users data is presented in Google Analytics. Note that users and session data are similar, but not identical.

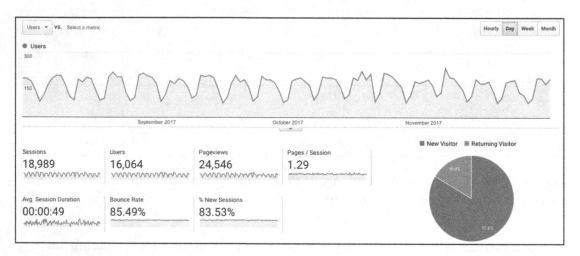

Figure 6-19. *A Users Report from Google Analytics*

Website Search Keyword Data

Your own website search tool can be a very rich source of data. Assuming you have search capability on your site, you can identify the more common search terms your visitors are entering in your tool when trying to find their desired information. You can also determine the number of times the users find the information and whether they go to the page provided in the search result.

This is very helpful for identifying what information your visitors are having trouble finding on your website. Knowing this can help you identify navigation or taxonomy changes that can help improve interaction on your site by making it easier for visitors to find the content they seek.

Figure 6-20 displays a Search Term Report from a website. The data includes the search terms entered by users and the number of times that term was searched for.

Search Term	Total Unique Searches ↓	Results Pageviews / Search	% Search Exits	% Search Refinements
	153 % of Total: 94.44% (162)	1.48 Avg for View: 1.49 (-0.68%)	30.07% Avg for View: 33.33% (-9.80%)	12.33% Avg for View: 11.98% (2.93%)
1. persona	4 (2.61%)	1.50	25.00%	0.00%
2. information architecture	3 (1.96%)	1.00	0.00%	33.33%
3. personas	3 (1.96%)	1.00	33.33%	0.00%
4. 5 second test	2 (1.31%)	5.00	0.00%	0.00%
5. analytics	2 (1.31%)	1.00	50.00%	0.00%
6. brainstorm	2 (1.31%)	1.00	100.00%	0.00%
7. click test	2 (1.31%)	1.00	0.00%	0.00%
8. labels information architecture	2 (1.31%)	2.50	0.00%	40.00%
9. Madina	2 (1.31%)	1.00	100.00%	0.00%
10. usability testing tools	2 (1.31%)	1.00	0.00%	0.00%

Figure 6-20. *An internal search tool report including the term and number of times it was searched*

Technical Data

Besides acquisition, conversion, and engagement, the technical information of the user experience on a site can also be an important source of behavioral UX data.

Knowing what devices, browsers, screen resolutions, and related technical information the visitors are using is important for evaluating how they are experiencing and interacting with the site. Often, using technical data to identify optimizations can improve the website experience for visitors without having to modify any of the content at all.

The following sections cover the more common types of technical data that can and should be evaluated.

Browser and OS

When evaluating the user experience of a website it is important to know what browsers and operating systems are most commonly used for viewing site pages.

The user experience of the site should work well for most browsers and operating systems, but the most common browser should have special attention to ensure that the experience for visitors using that browser is as optimized as possible.

It's dangerous to make optimization recommendations without first knowing how those changes will impact visitors using your most common browser.

Figure 6-21 shows a Browser and OS Report listing the top browsers by sessions. Note that Chrome is first, followed in a distant second by Safari, and then Firefox.

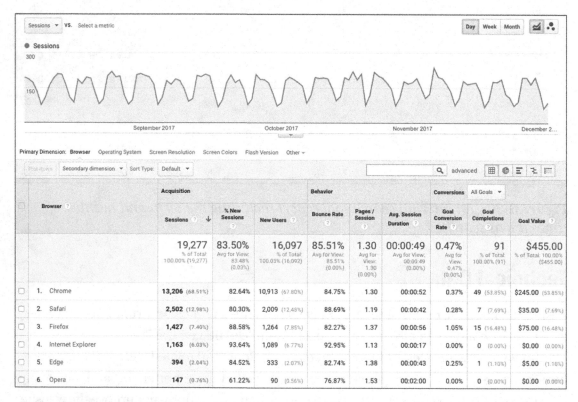

Figure 6-21. Browser and OS Report with a listing of top browsers by sessions

Mobile Devices

This data is helpful for identifying which mobile device or devices are the most commonly used when visiting your site. Knowing this is important for evaluating the experience your visitors are having based on their mobile device type.

Optimizing the site for all devices is important, but focusing on the most common devices and ensuring the experience is as good for them as possible is important for maximizing engagement.

Figure 6-22 displays a typical Devices Report in Google Analytics with a ranked listing of the top devices by sessions.

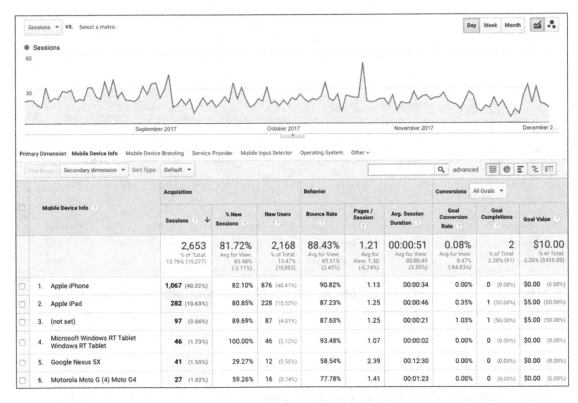

Figure 6-22. *This Devices Report displays a ranked listing of top devices by sessions*

Mobile Overview

The Mobile Overview Report in Google Analytics is helpful for identifying the activity on your site based on the type of device (mobile, desktop, or tablet).

Your site should work well across all device types, but ensuring you have the best possible experience for the most common device type is very important.

Figure 6-23 is an example of the Mobile Overview Report in Google Analytics displaying the top devices by sessions.

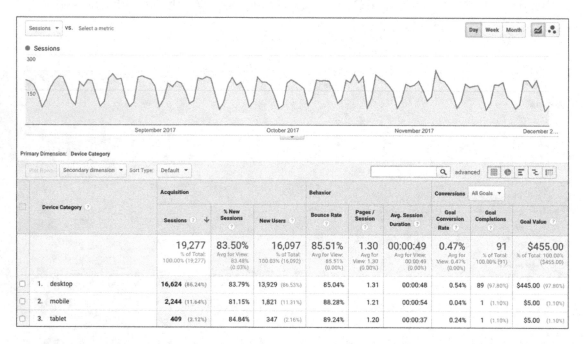

Figure 6-23. *The Mobile Overview Report in Google Analytics provides the summary of sessions by device*

Page Timings

The Page Timings Google Analytics report is helpful in identifying any pages that are experiencing longer page load times than the average for the site. This data can be used to hunt down poor performing pages based on page load, and then using other data sources to identify what may be causing the poor performance.

Figure 6-24 is an example of the Average Page Load Time Report displaying the average load time with a bar chart comparing average page load time to site average. This data is useful for identifying the pages causing poor load performance.

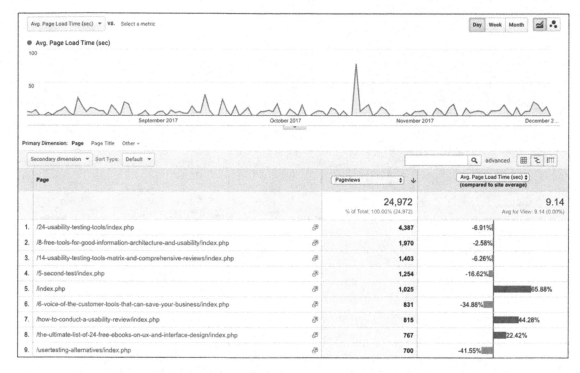

Figure 6-24. *The Average Page Load Time Report displays a bar chart showing page timings compared to the site average*

Screen Resolution

Another important metric to know and use is screen resolution. Knowing how your website looks and acts based on the top screen resolutions can offer insights into why certain user behaviors are or are not happening on the site.

Figure 6-25 is an example of a Screen Resolution Report from Google Analytics.

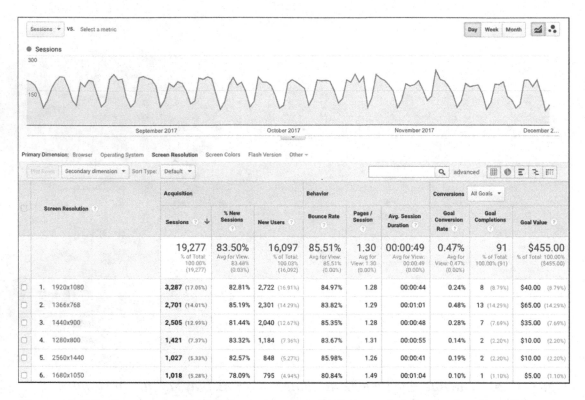

Figure 6-25. *Screen resolution is an important data point for evaluating the user experience of pages*

When evaluating how to optimize a website or particular page, it's important to ensure a user-friendly experience no matter what size screen the visitor may be using. But be sure to evaluate the page or pages using the most popular screen resolutions to ensure your user experience is as best as possible for that size screen.

Site Speed Overview

A critical element of your website's engagement is the speed with which your pages load. Slower load times have a direct impact on the number of users who tire of waiting for the page to load and end up leaving the site without ever seeing a page.

Tracking and optimizing page load speed is a very important aspect of optimizing the user experience of any site.

Figure 6-26 demonstrates how the Site Speed Overview Report provides details on average page load time, average server response time, and the average page download time, which are all useful for identifying how well the site pages are loading for users plus issues or opportunities for improving performance.

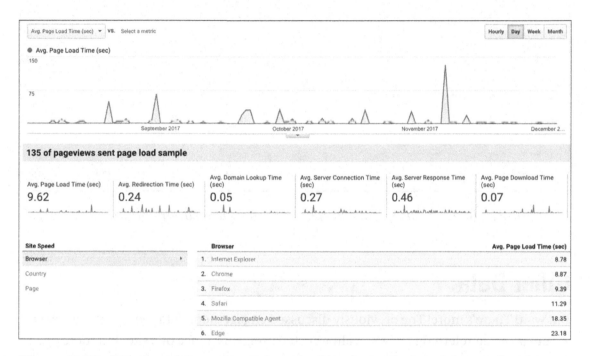

Figure 6-26. *Site Speed Overview Report in Google Analytics with page load performance metrics*

Speed Suggestions

The handy Suggestions Report in Google Analytics identifies by page what speed improvements Google recommends making to improve the load time of the page. This report connects with the Google Page Speed Insights Report that is available separately from Google.

Figure 6-27 provides an example of a Suggestions Report with links in the PageSpeed Suggestions column that connect with Google's PageSpeed Insights tool. This report lists specific recommendations for speed optimizations.

93

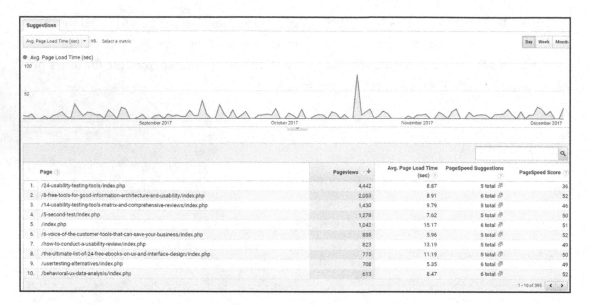

Figure 6-27. *Suggestions Report with links to specific recommendations for speed optimizations*

Other Data

But wait! There's more! The previously discussed engagement data represent the "usual suspects" often referred to when evaluating the user experience of a site. But there are literally hundreds upon hundreds of other data elements or variations of data elements that can be analyzed.

It's beyond the scope of this book to list them all out, but the above are excellent data sources you can use to start your analysis of the behavioral UX data.

Which data sources you ultimately choose depends on a number of factors including the questions you are trying to answer, the type of analysis needed, the type of website being analyzed (B2B, B2C, eCommerce), and the typical website engagement experience of your visitors.

Conclusion: Behavioral UX Data

Behavioral UX data is a very effective way to analyze the quantitative "what's happening" information on your website.

The four broad types of behavioral UX data include

- **Acquisition** (like PPC keyword data, etc.)

- **Conversion** (actions such as clicks, sign-ups, downloads, etc.)

- **Engagement** (such as bounce rate, time on page, etc.)

- **Technical** (visits by browser, screen resolution, etc.)

Although UX behavioral data is very helpful, this quantitative WHAT data is not enough. So what's missing?

Although you know WHAT is happening, you still don't know WHY it is happening.

Without this WHY information, any suggestions for optimizations are dangerous because you are forced to guess as to why the behaviors being reported in the behavioral UX data are occurring. Rather than guess, there's a much better way to find this all important WHY (or qualitative) data.

You must switch to qualitative UX research and usability testing to uncover the WHY for the observed WHAT behavior.

By conducting UX research or usability testing, you can begin to understand the WHY, which helps you make sense of the WHAT behavioral UX data you've already documented. Combining both of these data elements (quantitative and qualitative) provides you with a 360-degree view and a far more accurate set of information with which to make optimization recommendations.

I will cover qualitative UX research in the next chapter. Let's head on over there together now!

CHAPTER 7

UX and Usability Testing Data

A few years ago my client, the head of a website and app development company, hired me to conduct usability testing on a new app his team had spent the last ten months creating. The app was almost ready for delivery to the client. He was excited about his new app, which was designed to help people track their health.

The app was considered an MVP, meaning Minimum Viable Product. It had basic functionality including the ability to create a profile, log in, document various fitness activities and daily food intake items, and track all that against a health goal the user could set up.

He was very proud of the work he and his development team accomplished. He explained to me that he was confident the usability testing I was about to conduct would not find anything seriously wrong, but would most likely find just a few very minor issues.

I had already created the usability test protocol, and as always I wanted to test the test before conducting the actual moderated testing sessions. We decided to test the test on his new admin, who had recently been hired and had not experienced the new application yet.

The three of us sat in the CEO's office as I commenced the test of the test. The first task I wanted to test was setting up an account, which included setting up the account and entering basic health profile information, a mission-critical task for any app asking users to set up a profile.

"Can you please show me how you start using this new app?" I asked.

"Sure! I guess I have to log in or create an account or something," she said as she bent over the screen and started examining the first screen.

The login and profile creation process did not proceed smoothly. She struggled with the process of creating her account and profile. Several times she uttered "whoops" as she went back and forth between screens, trying to determine what went wrong as she

© W. Craig Tomlin 2018
W. C. Tomlin, *UX Optimization*, https://doi.org/10.1007/978-1-4842-3867-7_7

received error messages while trying to enter the information necessary to create her account and profile.

She paused.

"Well, I'm sort of stuck. I can't finish this. I don't know how to do this form," she said as she frowned, staring at the screen and clicking randomly.

"I don't know how to fix this. I think I'm stuck," she sighed, looking at the screen and continuing to tap randomly in the fields (which didn't work).

"So what would you do at this point?" I asked.

She looked at her new boss nervously before muttering, "Ummm, well, maybe go find some other app to use?"

Seeing her confusion, and her inability to easily fix what he considered a simple yet devastating set of usability errors, my client jumped up out of his chair as if his pants were on fire. He dashed out of his office and headed straight to the development team to let them know they had a big usability error that needed addressing right away.

The moral of this story? Two things:

1.   Supposedly simple stuff like creating an account and profile can be made difficult and confusing to users if the process does not match their mental map.

2.   Even a test of a test with just one user can surface important usability issues that can escape even the smartest design and development team.

This is an example of using UX research to find and uncover issues. In addition, UX research and usability testing analysis provides the qualitative WHY data for website or app interaction. By coupling this data with the "what's happening" quantitative behavioral UX data discussed in the prior chapter, you are able to make far more informed optimization recommendations for websites.

In this chapter, I will provide an overview of

-   What UX research and usability testing is (and isn't)

-   The types of data available

-   How to analyze this data

-   Examples of how this data is used

UX Research and Usability Testing Analysis Goal

The goal for UX research and usability testing analysis is to obtain qualitative WHY data for the "what's happening" data coming from the behavioral UX data analysis.

Your method for conducting UX research and usability testing is to gather qualitative data by observing actual people who match the Personas as they conduct critical tasks on your website.

Questions you want to answer by conducting this qualitative analysis include

- What is **working** for website visitors?

- What is **not working** for them?

- What **confuses them** or causes them concerns?

- Are their **expectations** met for the experience?

- If expectations are not met, why not?

You will use this data to evaluate and identify patterns of task-flow errors. In addition, the qualitative component can help in identifying the potential WHY for the task-flow issues that may be uncovered.

Types of UX Research and Usability Testing Methods

There are many UX research and usability testing methods you can use to capture qualitative data:

- **Moderated Usability Test**: Richest source of qualitative data, but resource intensive

- **Unmoderated Usability Test**: Good source of qualitative data, and less resource intensive

- **5-Second Test**: Provides data for how well a page communicates with visitors

- **Card Sort**: Excellent for evaluating information architecture questions

- **Click Test**: Identifies navigation and taxonomy issues

- **Eye Tracking**: Identifies good and bad elements that may be helping or hurting attention

- **Preference Test**: Can add the WHY data for preferences among several choices

- **Question Test**: Useful for identifying task-flow and page communication problems

- **Etc. Etc. Etc.**: Besides the main types above, there are many other types of UX research and usability testing qualitative methods and tools available. Which ones you choose depends on a number of factors including the type of analysis needed, the type of Personas being tested, the type of website (B2B, B2C, eCommerce, etc.), and the detail needed for analysis.

Let's review each one with a bit more detail.

UX Research and Usability Testing Methods

Let's start with what is arguably the richest source of qualitative data, yet most resource intensive: moderated usability testing.

Moderated Usability Testing

Moderated in-person or moderated remote usability testing provides the richest and most detailed form of usability testing data. This method requires the moderator to be present either in person or through a remote live connection via computer with the participant. The moderator reads each task to the participant, who uses the "think aloud" method of talking while interacting with the website or app. Think aloud means the participant explains what they are thinking about as they move through the website and their tasks. The moderator can follow up or probe to uncover additional information from the participant as they go through the tasks. Figure 7-1 is an example of a moderated in-person usability test with the participant's screen above and the video image of the participant as she is conducting the test.

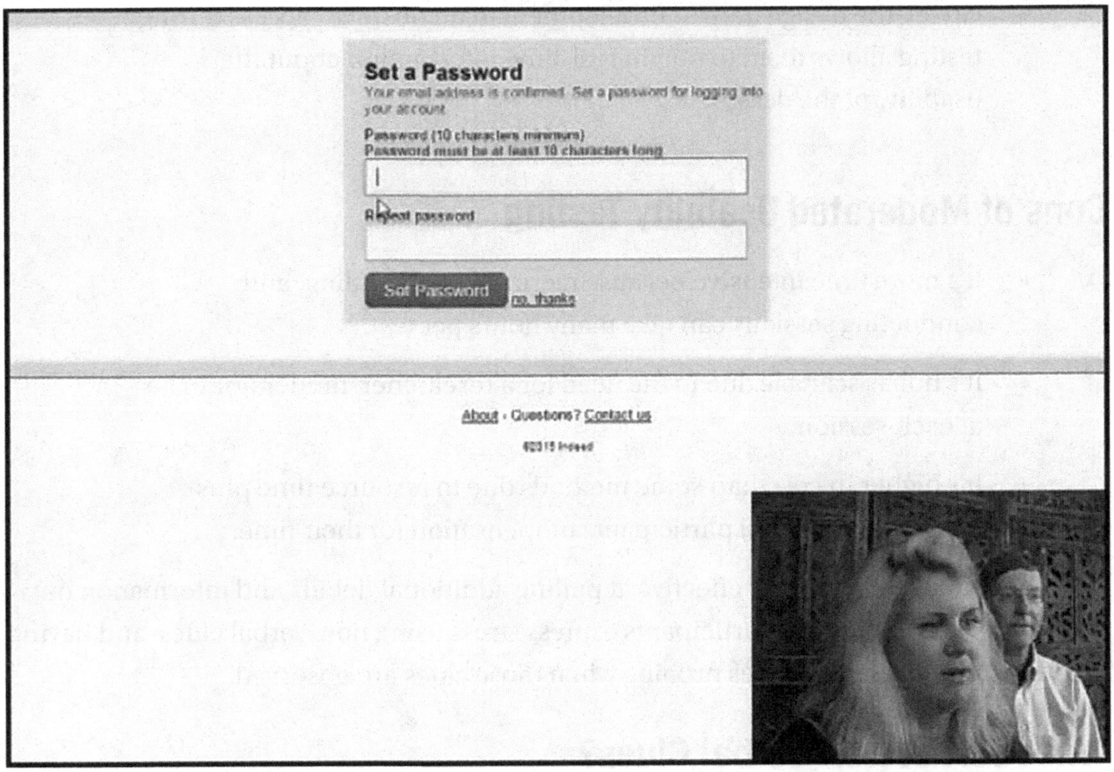

Figure 7-1. *Screenshot of an in-person moderated usability test*

Typically moderated usability tests are recorded (the screen and the participant's voice and face), so the research team can refer back to the session as needed. Often highlight video reels combining several short snippets from several tests are created to demonstrate where usability issues are occurring and what the participants are saying or thinking as those issues are happening.

Because it's the critical task being tested, and not the participant, moderated tests only require from five to ten participants to find key usability issues.

Pros of Moderated Usability Testing

- It's the richest form of usability testing due to moderator's ability to probe and follow up with the participant.

- Actual users interacting with the site or design provide a "real world" view into what parts of the design are working well or may need improvement.

- Giving the design and/or development team observer access during testing allows them to obtain real-time information about the usability of the design.

Cons of Moderated Usability Testing

- It's more time intensive because recruiting, scheduling, and conducting sessions can take many hours per test.

- It's not as scalable due to the need for a researcher/moderator to be at each session.

- It's higher in cost than some methods due to resource time plus recruiting costs and participant compensation for their time.

Moderated testing is very effective at pulling additional details and information out of participants. Frequently participants express stress using non-verbal clues, and having the moderator present enables probing when those clues are observed.

What Sort of Non-Verbal Clues?

Non-verbal stress clues that may indicate the participant is having difficulty (i.e., cognitive load) include

- Raised eyebrows

- Sighs

- Perplexed looks

- Sitting back in their chair

- Staring at the screen

- Utterances of surprise such as "oh," "um," etc.

- Rapid movements of the mouse or screen without clicking

When to Use Moderated Usability Testing

Moderated usability testing works well in situations where a moderator is needed for probing or help moving participants from test to test. These situations often occur when testing low fidelity, early prototypes or wireframes, in which some explaining or help is needed to move participants from test to test. Another situation well-suited for moderated testing is when probing and follow-up are required to uncover additional details as to why certain issues are occurring.

Several vendors offer remote moderated usability testing. Some of the services include Usabili.me, UserTesting, and Validately.

Unmoderated Usability Testing

As with moderated usability tests, unmoderated usability testing is a powerful way to capture the all-important WHY qualitative data.

The key difference is that in unmoderated usability testing the moderator is not present with the participant when they conduct the test. Instead, the researcher provides task instructions for each task they want tested, and the participant is recorded as they try to accomplish the task.

The participant uses the think-aloud method, which provides the WHY information as they move through the process flow. Figure 7-2 is an example of an unmoderated usability test in action.

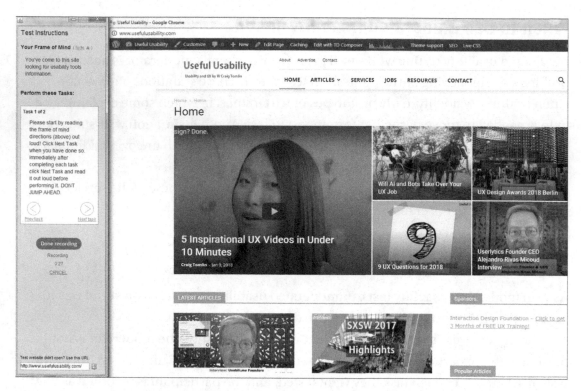

Figure 7-2. *An unmoderated usability test in action from the participants perspective using TryMyUI. Resulting recordings can be reviewed for analysis after the test has been completed.*

Because a moderator is not present for each test, unmoderated testing can be conducted much faster than moderated testing, requiring lower effort, time, and cost. They can also be conducted at scale, with 5, 10, 20, or even more tests all happening at the exact same time. And because they can be done remotely, unmoderated usability testing can be conducted anywhere, at any time, and with any participants who have the technology to conduct the test.

So how do you obtain that WHY information if a moderator is not present?

The answer lies in how carefully the unmoderated test is set up. By focusing on specific task-based questions and reminding the participant to think aloud as they conduct the task, the WHY information can be captured. Recording the session allows the UX researcher to retrospectively observe the session and analyze it for the WHY of any performance issues.

If set up correctly, unmoderated usability testing can be almost as effective as moderated testing. Almost.

Pros of Unmoderated Usability Testing

- It's a scalable and fast solution that enables multiple usability tests to be conducted at the same time.

- Actual users interacting with the site or design provide a "real world" view into what parts of the design are working well or may need improvement.

- Remote unmoderated testing can be conducted with real users anytime, anywhere, and with anyone that has access to the testing tools.

Cons of Unmoderated Usability Testing

- Because there is no moderator, the ability to probe and follow up with verbal or nonverbal clues is absent.

- Some panels may have a few "professional" testers who may not reflect the typical user that may visit a website.

- Without a moderator being present, the script for walking participants through the test is critical, and it can accidently include biases if not prepared very carefully.

When to Use Unmoderated Usability Testing

Unmoderated usability testing works well if testing is being conducted on live sites or high-fidelity working prototypes. Testing on critical tasks that are easily identified as having a success or failure condition also lend themselves to unmoderated testing. Because they are scalable and fast, unmoderated usability testing is useful for quick tests in which results are needed in hours, or at most a day.

A variety of vendors offer remote unmoderated usability testing services, including TryMyUI, Userlytics, and UserZoom. Most of these services offer either a panel of participants you can select from using demographic or screener type questions, and many of them allow you to use your own participants to conduct the test.

5-Second Test

A 5-Second Test is a test of the communication ability of a webpage. The concept is simple: participants view an image of webpage for 5 seconds. At the end of the 5 seconds, the image is removed and the participant is asked questions about what they've just seen.

5-Second Tests are extremely helpful for evaluating how well a website communicates three critical pieces of information to visitors:

- Who are you? (the brand/owner of the website)

- What do you offer? (summary of product or services offered)

- Why should they care? (what's in it for the visitor)

Why 5 seconds?

Because according to research conducted by Lindgaard et. al.[1] and other research studies, most visitors actually make their minds up about a website's quality within only 50 milliseconds.

Yes, that's MILLISECONDS.

And as demonstrated in our earlier discussion of quantitative data, the Session Duration Report for websites will typically show that the vast majority of visits are actually LESS THAN 10 SECONDS.

Yikes!

So 5 seconds has become something of an industry standard for evaluating how well a website communicates those all important pieces of information to visitors.

As with unmoderated usability testing, the devil is in the details in setting up an unbiased 5-Second Test. It is very important that questions are carefully worded to reduce bias and are focused on answering the three critical communication elements of who you are, what you do, and why they should care (how it helps them).

In the 5-Second Test results shown in Figure 7-3, 63 percent of the participants were unable to determine the product or service offered by this website. That's not good because this was a test of the website's home page!

Ouch.

---

[1]Gitte Lindgaard, Gary Fernandes, Cathy Dudek, and J. Brown. "Attention web designers: You have 50 milliseconds to make a good first impression!" Behaviour & Information Technology, Volume 25, Issue 2. 2006. Taylor & Francis Online. Pages 115-126. Published online, March 4, 2011. www.tandfonline.com/doi/abs/10.1080/01449290500330448. Accessed December 7, 2017.

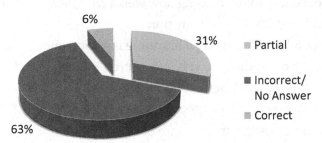

Figure 7-3. *5-Second Test results for the home page of a website*

The good news is a 5-Second Test is a very powerful tool for identifying not only WHAT is not working, but WHY it is not working. That's because you can include probing questions at the end of the test to help uncover the WHY qualitative data.

There are online tools that help make conducting a 5-Second Test very fast and very scalable. Testing can be set up and run in a day, or hours, and the results can include dozens, hundreds, or potentially thousands of data points.

Pros of 5-Second Testing

- It's a scalable and fast solution that enables obtaining dozens or hundreds of results in as little as a day.

- Actual users provide the information and results.

- Follow-up questions can probe to learn what elements are or are not attracting attention and why.

Cons of 5-Second Testing

- Testing is limited to a unique page, so the experience of "surfing" a website by viewing multiple pages to learn about a product is not present.

- The image presented is static, so any interactivity that might normally be present in a webpage with moving visuals is not present.

- The questions and the order they are asked is critical and can accidently include biases if not thought out very carefully.

When to Use 5-Second Testing

A 5-Second Test is excellent for evaluating how well a webpage is communicating critical pieces of information and can be used any time.

The test is very good for identifying if branding is strong enough to be retained by a visitor, if the messaging of products or services offered is clear, and if visitors understand why the products or services may benefit them.

5-Second Tests are important whenever a website is being redesigned or new pages that change formatting or visual structure are being developed.

A vendor that offers remote 5-Second Tests is FiveSecondTest.com (via the parent, UsabilityHub).

Card Sort

A card sort is a UX research method in which participants organize items into groups or buckets of similar information. Card sorts are a very useful tool for determining the information architecture, labeling, and navigation of a website. The data from a card sort enables a comparison of the existing website information architecture and navigation compared to the mental map of the users who visit your site.

Card sorts were named as such because originally 3x5 index cards were used to conduct the sort. Participants would sit at a table with a set of 3x5 cards randomly distributed around the top. Each card had a single term on it; typically that term came from the navigation terms used on the website.

Participants were asked to sort the cards into piles, each pile being a group of like items. Often, once the participant completed making their piles, they were asked to add a card on top of the pile with a term that would represent the label for that pile.

In addition to the labeled cards, several blank cards were provided so the participant could write their own labels for piles if they felt they needed to.

Today's card sort testing uses online virtual cards in place of the 3x5 index cards, but uses the same methodology for uncovering the bucketing of information from the user's perspective. See Figure 7-4.

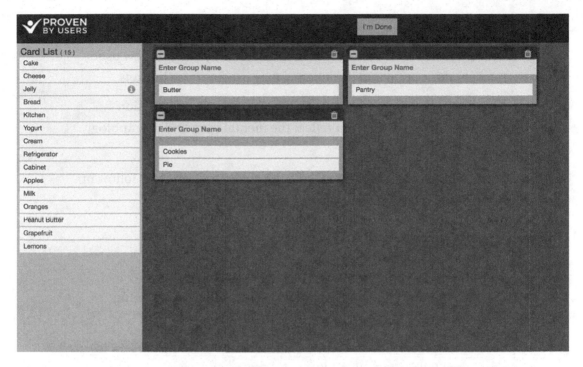

Figure 7-4. *Card sorting example with cards to be sorted on the left and groups the participant is creating on the right*

Information architecture and navigation labeling studies can be conducted extremely quickly and with participants anywhere in the world, thanks to the Internet.

There are three types of card sorts:

- **Closed**: Groups are created and labeled by the test creator.

- **Open**: Groups are created or labeled by the participants.

- **Hybrid**: Some groups are created by the test creator, but they can be modified or added to by the participant.

Pros of Card Sorts

- Online card sorts are a very fast and scalable way to learn how users bucket information and the labeling they use for those buckets.

- Conducting sessions with actual users ensures the information architecture and navigation reflect the mental map and terminology of users.

- Many online card sort tools include the ability to probe and follow up by enabling questions and surveys at the end of a test.

Cons of Card Sorts

- As with much of UX research, card sorts can provide bad data if the participants do not match typical users for the system being investigated.

- Online card sorts lack the ability of the moderator being present to probe and follow up with additional questions.

- Card sorts are useful for labeling items that are known to users. New items or those not used often by users may cause issues with test results.

When to Use Card Sorting

Card sorts can be used at any time during a website development or redesign project. Best results happen when the card sort is conducted prior to website development because results can be used to create the information architecture and navigation schema. Card sorts are also effective whenever a website is being updated because results can help guide optimization of the navigation and information architecture.

Vendors that offer online card sorting tools include Provenbyusers.com, Simple Card Sort, UsabilityTools, UserZoom, and UsabiliTest.

Click Test

Click tests, sometimes known as first click tests, help identify navigation and taxonomy issues with a website or app.

Click tests work by presenting the participant with an image of a website or app. The participant is asked to click where they think they would complete a particular task. The location on the image where the participants click is recorded, and heat maps or related data visualizations can be produced showing where the highest number of clicks occurred.

Examples of click tests include asking the participant where they would click to find help, or sign-up for a service, or find information about xyz product.

Because click tests can be done very quickly, and with any design image like wireframes or mockups, they are a very useful way to check navigation and labeling for new designs that are not yet live.

Most click test tools record where on the screen the participant clicked and how long it took for the participant to click. This data is typically presented in heatmaps where high concentrations of clicks are in hotter colors like reds or yellows and fewer concentrations of clicks are in cooler colors like greens or blues.

By adding questions after participants complete the click test, important WHY qualitative data can be obtained to help inform recommendations for optimizations.

Figure 7-5 is an example of a click test in action. The participant is asked to complete a task. The participant clicks where they think they would complete the action.

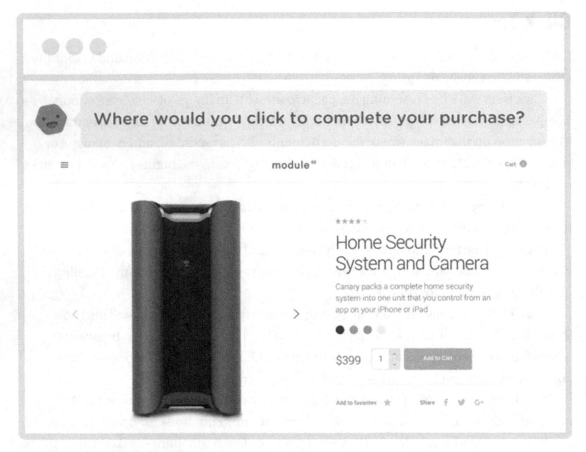

Figure 7-5. *Click test with task-based question. The participant clicks where they think they would complete the task.*

Figure 7-6 shows the results of a click test. The heatmap provides a visual representation of locations that received a higher number of clicks. This data is helpful for evaluating the task-flow and identifying any issues with users navigating the system and completing tasks.

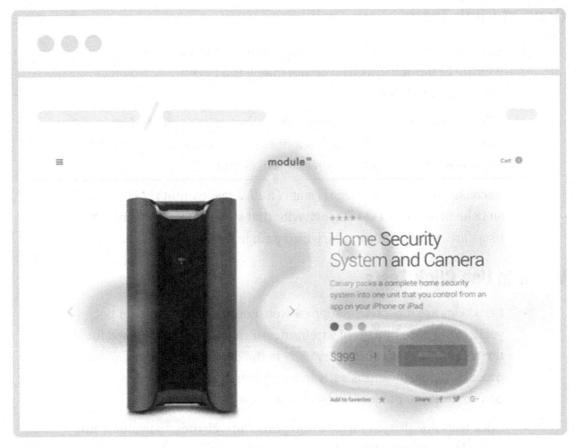

Figure 7-6. *Heatmap visual representation of the location of participant clicks in response to the question*

Pros of Click Tests

- Click tests enable evaluations of where participants do and do not click when they are trying to accomplish tasks or navigate.

- Click testing is simple to set up, scalable, and fast, providing results in just a few hours.

- Click tests are a good way to validate a new design prior to it going live.

113

Cons of Click Tests

- Click tests are subject to bias. Participants who already know a website or participants who do not match the Persona of a typical user can skew results.

- Click tests are typically used for a single screen or at most a few screens deep, which means that longer task-flows requiring multiple screens may not lend themselves to a click test.

- Because the click test image is static, a click test cannot replicate the website browsing or surfing behavior that sometimes occurs when users are trying to determine which path to take on a website.

When to Use Click Tests

- Click tests are most useful for new websites or designs that are not yet live. Because images of wireframes or mock-ups can be used for the test, testing can occur very early in the design and development process. Using click tests early in the process can help improve conversion well before a website or app ever goes live.

- Click tests are also useful for testing existing websites, to evaluate not only places people should be clicking, but also places they should not be clicking.

- Click testing can reveal web design treatments that are causing people to click designs that are not actually links. Knowing this can help designers improve the designs to reduce the number of wasted clicks and improve the number of good clicks.

Vendors that offer online click testing tools include UsabilityHub and UserZoom.

Eye Tracking

Eye tracking is the process of measuring the gaze points of a participant as they are viewing a webpage or other visual object. It's beyond the scope of this book to get into the technical details of how eye tracking works, but a very high-level overview is helpful.

Eye tracking uses hardware and software to monitor and track movements and fixations of a participant's eyes. A fixation is when the eye momentarily stops (for only hundreds of microseconds) to gather visual information. By gathering fixations and related information like saccades (the rapid movement of the eye from one fixation to the next) and scan paths (a series of saccades and fixations), the gaze points of a person's eyes can be recorded and presented as heatmaps. With eye tracking heatmaps, greater areas of fixations have hotter red colors and lower areas of fixations have cooler blue or green colors. Areas with no fixations have no colors.

The old, kludgy, hard-to-use helmets that contained the hardware to conduct eye tracking have given way to sophisticated glasses, and now no wearable hardware at all, to track eye movements. Eye tracking is effective for obtaining data on where people look, and sometimes just as importantly where people do NOT look. Figure 7-7 demonstrates the results of an eye tracking study using heatmaps to display areas of greater and lesser fixations.

Figure 7-7. *Eye tracking heatmap from Feng-Gui of a web page, with potential higher fixations areas in red/yellows*

Pros of Eye Tracking

- It provides data on what elements on a page are or are not attracting attention.

- The data is useful for understanding the WHY of people either clicking or not clicking elements on a page.

- Follow-up questions can probe to learn what elements are or are not attracting attention and why.

Cons of Eye Tracking

- Testing requires special hardware or software required to collect data.

- The act of observing objects, or not observing them, does not necessarily imply an action will occur, so the data may not always correlate with website activity.

- The use of special equipment or software may skew results by removing participants from a more normal environment.

When to Use Eye Tracking

- Eye tracking data can help you evaluate what elements may or may not be attracting several types of attention:
  - "Good" attention points
  - "Neutral" attention points
  - "Bad" attention points

Graphics and visual design on a website can be powerful tools for capturing a visitor's attention. This is extremely helpful when you want your visitors to know important information or direct their attention to a call to action such as a "Buy Now" button. These can be considered "good" attention points because the attention is helping the user accomplish goals.

However, graphics and visual design can also accidently pull a visitor's attention AWAY from the very information or calls to action you were hoping your visitors would see. Those items can be considered "bad" attention points. An example would be a graphic image that looks like a clickable button, but is actually not clickable.

As the example heatmap of eye tracking data results in Figure 7-7 demonstrates, areas where there are more fixations are colored green to yellow to red to indicate lower to higher amounts of fixations. Areas of no fixations have no color.

This data can help inform you about what elements on a page are attracting or not attracting attention. It can also inform you about objects that should NOT be attracting attention, but are. They can be labeled as "bad" attention sources because they may be harming the user experience by pulling visitors away from focusing on the elements they seek.

So in the example in Figure 7-7, the hot spot areas of the search bar, shopping cart, and product images are all good attention areas. But perhaps some of the text in the middle, which is not clickable, may be pulling attention away from the all-important products and could be considered neutral or potentially bad attention areas.

There are multiple vendors that offer eye tracking tools including EyeLink, Feng-Gui (automated tool), Gazepoint, SmartEye, and Tobii.

Preference Test

A preference test is a qualitative tool that can help determine which of several design variations participants prefer. Preference tests are helpful for providing user insight as part of a design process.

Preference tests work by showing several versions of a design to a participant who then selects which variation they prefer. After their choice is made, the participant typically completes several open-ended questions as follow-ups. These questions can include the degree to which they prefer their choice and why they picked that choice. This is where the all-important WHY qualitative data comes from.

Online preference testing is a very fast and effective way to capture user-centered data because the tool is available to anyone anywhere in the world who chooses to participate in the test.

Figures 7-8, 7-9, and 7-10 demonstrate what participants experience when they conduct a preference test using UsabilityHub.

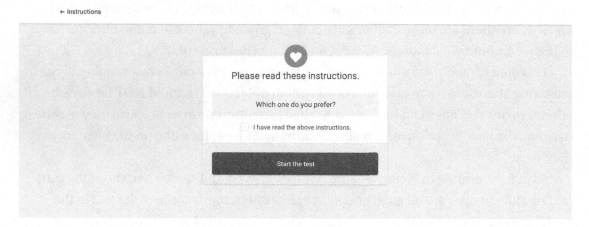

Figure 7-8. *The instruction page for the preference test*

Figure 7-9. *The choices for the preference test; participants click their choice*

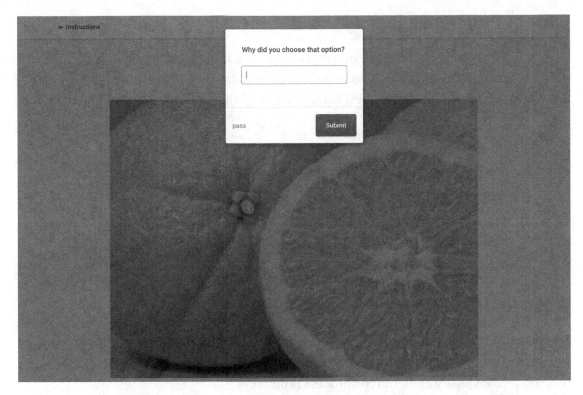

Figure 7-10. *The follow-up WHY question about the participant's choice*

The first screen (Figure 7-8) provides the instructions for the test. The second screen (Figure 7-9) appears after the instruction screen and displays the two choices presented to the participant. The participant clicks on their preferred choice. After selecting their preferred choice, the participants are presented with the third screen (Figure 7-10) where they respond to follow-up questions that probe the WHY of the choice.

Pros of Preference Testing

- Preference testing is useful for determining which variation of a design users prefer, and why they prefer that design.

- Because testing is conducted online, participants can be recruited from anywhere in the world where there is Internet access and at any time.

- Preference tests are a helpful way to resolve discussions among a design team on which design variation would be the better choice.

119

Cons of Preference Testing

- Preference tests can provide inaccurate data if the participants do not match the Personas, if there are not enough participants to provide accurate data, or if the test itself is not carefully developed to exclude bias.

- Preference tests are based on which design a participant prefers, but just because the design is preferred doesn't necessarily mean the design will perform better.

- Preference tests should include the additional follow-up questions; otherwise the choices made have no context for why the choice was made and thus no qualitative data.

When to Use Preference Testing

The best time to use preference testing is when there is a need to determine which of several iterations users prefer. This can occur at any time in a website design or redesign project, but obviously earlier in the process is better.

Multiple vendors offer online question test tools including Verify (Helio) and UsabilityHub.

Question Test

Question tests are useful for identifying task-flow and page communication problems. In a question test, participants are provided with an image of a webpage and asked questions about the page. Because questions tests are online, they can be conducted anywhere and at any time, as long as the participant has access to a computer and the Internet.

Unlike a 5-Second Test, with a question test participants have as long as they wish to review the image and answer the questions. Results for a question test include the raw response data and often visual representations of the data, often using word clouds.

Figure 7-11 is an example of a question test in which the participant has as much time as necessary to view the image and answer the question.

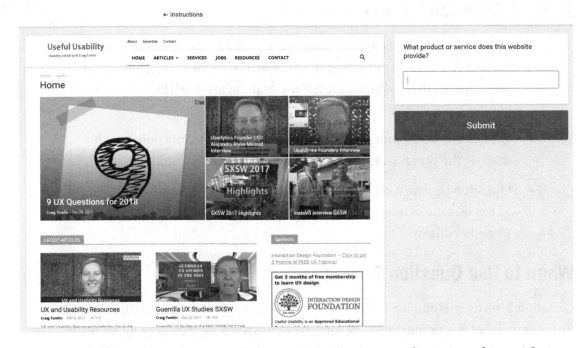

Figure 7-11. *In a question test, the participant has as much time as they wish to review the image on the left in order to answer the question on the right*

What Sort of Questions Can Be Asked in a Question Test?

Typically the questions are focused on learning how well the layout or content in a page is working. Examples might include trying to evaluate if the testers understand the concept in copy on the page, if they know where to go to complete certain tasks, and what they would expect to happen next if they were to take an action on a page.

Pros of Question Testing

- Question tests can be conducted very early in the design process, using images of page mockups or even wireframes.

- Because testing is conducted online, participants can be recruited from anywhere in the world where there is Internet access and at any time.

- Question tests are useful for evaluating how well content or navigation on a page is communicating to users.

Cons of Question Testing

- Like preference tests, question tests can provide inaccurate data if the participants do not match the Personas, if there are not enough participants to provide accurate data, or if the test itself is not carefully developed to exclude bias.

- Question tests cannot solve design issues; they can only point out if there is the potential for an issue to be present.

- Question tests that do not provide WHY questions as follow-ups to the questions may lack the necessary qualitative data needed to provide context.

When to Use Question Testing

Question tests are helpful for analyzing how well a page is communicating through its content, labeling, or functions. As such, question tests are excellent for website redesign projects when page task-flow, labeling, or content needs to be evaluated and any identified issues prioritized. Question testing is also very helpful for new websites or apps, and can be used with mock-ups or even wireframes.

Multiple vendors offer online preference testing tools, such as UsabilityHub.

Other UX and Usability Testing Analysis Tools

There are plenty of other UX and usability testing analysis tools that you can use to help you obtain the WHY qualitative data you need. Many are based around capturing input from people who are interacting with the site, including the types covered in the following sections.

Contextual Inquiries

Sometimes known as ride-alongs or field visits, contextual inquiries are visits to a user's environment to observe and learn how they typically interact with a system.

Product and design teams typically use this approach to learn how the user interacts with a system and what parts of the user's existing system do and do not work well.

The feedback gathered from a contextual inquiry is often rich in detail but is also unstructured data. Unstructured data in this case means the notes, observations, and other random information compiled during the contextual inquiry are not easily parsed and thus need human intuition to evaluate and analyze.

To that end, the notes, details, and related observations have to be carefully vetted to ensure the data reflects actual usage and not just opinions or guesses as to how the system is being used.

Contextual inquiries are not used for usability testing tasks on websites or apps. However, the data from contextual inquiries can be very useful for evaluating what tasks may need to be tested on a site or app.

Diary Studies

Diary studies are a longitudinal research method often used for market research and other ethnographic studies. A longitudinal study is one that occurs over a relatively long period of time (days, weeks, months, etc.).

The goal of a diary study is to have the participant record and document their activity with systems and/or their thoughts or feelings about a system over time.

The information from a diary study is then consolidated into key findings that can be used for a variety of purposes, but often are used in market research or product development analysis.

Because diary studies are self-reported by the participants and are not tests of specific usability tasks, diary studies are not used for usability testing. However, data from diary studies can help inform or inspire research directions into what should be tested on websites or apps.

Focus Groups

A focus group is a small group of people who are asked to come together in a real-life setting or virtual room with a moderator. The goal of a focus group is to have the participants provide insights, ideation, and feedback in a group-think environment.

Focus groups can be useful tools for obtaining direct user feedback on their opinions, attitudes, and beliefs. Because the moderator is present, focus groups have the added advantage of the moderator being able to probe and follow up on participant comments as necessary.

Focus groups are not used for usability testing research because what users say they do vs. what they actually do is often quite different. However, data coming from focus groups can trigger additional UX research studies and usability testing based on insights gathered from participants in the focus group.

Surveys

Surveys are not typically considered qualitative tools. However, adding qualitative questions that seek to uncover the WHY behind opinions, beliefs, and attitudes can be a source of qualitative data derived from surveys.

Surveys require much larger numbers of completions vs. many other UX research methodologies. This is due to the need to reduce the potential for the data to have errors based on being skewed by having too few data points.

Statistical significance and error rate are two important metrics that help determine whether the data in a survey is reliable or not.

Surveys are not used as a tool for usability testing research due to several factors: one being what people say they do is often not what they actually do, and another being the inability to observe users interacting with specific tasks on a system. However, the data coming from surveys can be used to help determine potential usability testing and UX research studies.

Some UX research studies use surveys at the end of a test to further probe and provide additional insights into the results of the UX research study.

Voice of the Customer (VoC) Tools

VoC tools include feedback forms, questionnaires, or ratings on websites or apps that enable visitors to provide input on the quality of the page. This kind of feedback is helpful but can also be somewhat skewed toward unique one-off issues that have nothing to do with the website or app.

Figure 7-12 is an example of a VoC feedback form. The first question, "Was this article helpful?," probes whether the user was able to use the provided content to help them answer their questions.

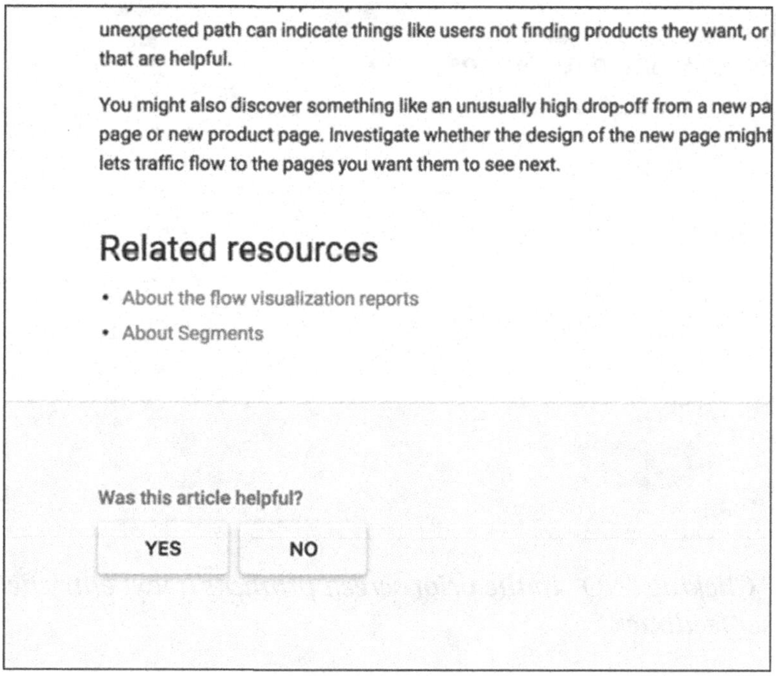

unexpected path can indicate things like users not finding products they want, or that are helpful.

You might also discover something like an unusually high drop-off from a new pa page or new product page. Investigate whether the design of the new page might lets traffic flow to the pages you want them to see next.

Related resources

- About the flow visualization reports
- About Segments

Was this article helpful?

| YES | NO |

Figure 7-12. *A website VoC feedback form to capture user data on quality of content*

In Figure 7-12, if the user clicks the NO button, they are offered a text entry field to provide additional feedback on how the content can be made better.

Figure 7-13 is a screenshot of the text entry field for additional user feedback. This is a classic example of a VoC website tool in action.

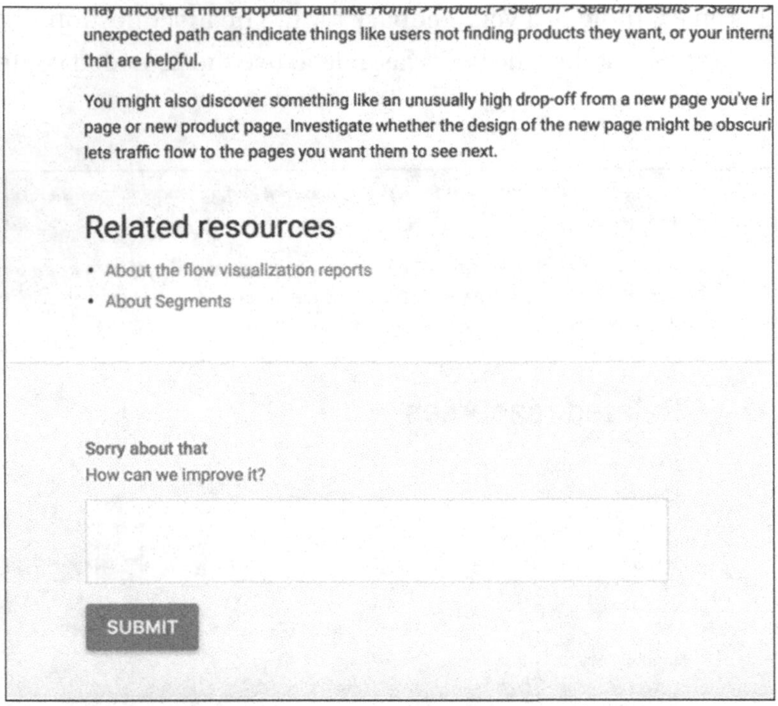

Figure 7-13. *Clicking "NO" in the prior screen prompts a text entry field for additional user feedback*

VoC tools are not used for usability testing because they do not provide the ability to have the tester conduct tasks. However, the data coming from VoC tools can help inform usability testing research studies.

Conclusion: UX and Usability Testing Analysis

There are other tools available to gather qualitative data, but the ones listed above are typically the most used for UX research and are the best for capturing and analyzing various forms of qualitative data.

The examples above should give you a good sense of what qualitative data is possible to capture and how that data can be used.

You can use a variety of UX and usability testing tools to gather qualitative WHY information about website activity. These tools can include the following:

- **Moderated Usability Test**: Richest source of qualitative data, but resource intensive

- **Unmoderated Usability Test**: Good source of qualitative data, and less resource intensive

- **5-Second Test**: Provides data for how well a page communicates with visitors

- **Card Sort**: Excellent for evaluating information architecture questions

- **Click Test**: Identifies navigation and taxonomy issues

- **Eye Tracking**: Identifies good and bad elements that may be helping or hurting attention

- **Preference Test**: Can add the WHY to preferences among several choices

- **Question Test**: Useful for identifying task-flow and page communication problems

The goal of using these tests is to obtain the all-important WHY qualitative data coming from website activity. By analyzing this data, you will have a much better picture of what is occurring on your website and possible reasons why it is occurring.

By combining the quantitative WHAT behavioral data you captured earlier with this qualitative WHY data, you now get a much more complete, 360-degree view of what is happening on the website and why it is happening. This gives you a much better set of data with which to make website optimization recommendations.

So what's next?

Once you've obtained your behavioral UX data and your UX and usability testing data, it is time to combine them for analysis and recommendations. And that, luckily enough, is the subject of the next chapter!

CHAPTER 8

Putting It All Together: Behavioral UX Data Analysis and Recommendations

Have you ever seen one of those pattern eye tests that asks if you can read the words, something like Figure 8-1?

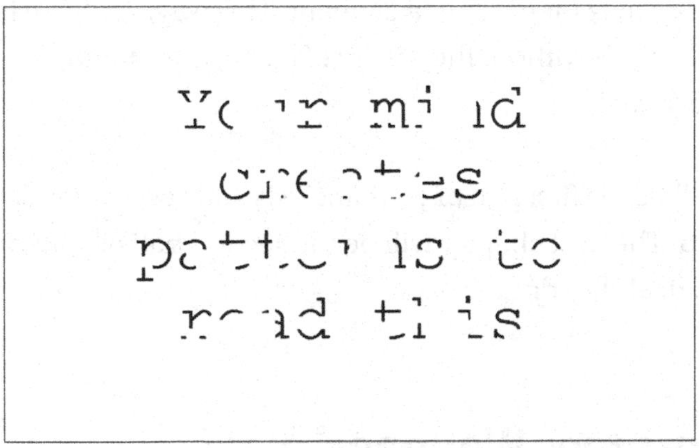

Figure 8-1. *Example of a reading test using pattern recognition*

Most likely, you were able to read the text in the pattern test, even though the letters have large chunks of visual data missing. It says

Your mind creates patterns to read this

© W. Craig Tomlin 2018
W. C. Tomlin, *UX Optimization*, https://doi.org/10.1007/978-1-4842-3867-7_8

We can read it because our brains are wired to look for patterns and to fill in any blanks to create the resulting shape.

Just like the example eye test, your goal in putting it all together is to look for patterns among diverse pieces of data. The patterns you will look for are created by combining everything you have acquired in your research to this point. So far, you have

- Identified the Persona or Personas

- Analyzed behavioral UX data

- Evaluated UX research and usability testing data

You will start looking for patterns across each of these data sources. Using your ability to combine elements into a consistent pattern, you will identify issues and thus be able to make recommendations for optimizations that can then be vetted using A/B testing.

Case Study: eCommerce Website Optimization

The best way to help you understand how to do all this is to walk through a case study. But I won't just give you the answers to the data analysis. That would be too easy!

No! Instead, you'll have to put away your books, sharpen your number two pencils, and do a little practice quiz for each data element. This way, you'll be far more ready to conduct your own analysis without the luxury of having me standing over your shoulder giving you all the answers!

Fun! Right?

This chapter will be the first half of your analysis and recommendations: **behavioral UX data**. The next chapter will be the second half of your analysis and recommendations: **usability data**.

So let's begin!

Shop.MyEvergreenWellness

My client, the marketing agency Marketing In Color, is a skilled marketing firm with a long history and robust portfolio demonstrating sophisticated brand strategy, creative, video, websites, and much more. Development of the site they asked me to review was on a very tight timeline that required moving from initial concept to launch in approximately four weeks.

Conceding that the condensed timeline would mean certain elements would have to be prioritized over others meant they would not have the opportunity to refine the site before launch. Their strategy was to get to market quickly and record early analytics to establish a performance benchmark. They could then use this as the foundation from which to evolve and improve the site over time. This is the point at which they engaged me.

The site they asked me to evaluate is `Shop.MyEvergreenWellness.com` (Figure 8-2). It is an eCommerce website focused on helping active agers to live a healthy lifestyle though proper nutrition, exercise, and mindset. The website includes paid and free programs designed to help them achieve their goals and live their best life.

Figure 8-2. *The Shop.MyEvergreenWellness.com site*

So let's say they call you asking for your expertise in taking the site to the next level. They are looking for ways to increase the conversion of visitors to buyers. They are also interested in increasing the conversion in the buy-flow by reducing the abandonment rate in the cart.

They have tried a variety of techniques to increase conversion, including conducting A/B testing in and out of the buy-flow. Conversion has moved up, but they are looking for you to help them improve conversion even more.

The Shop team (I'll shorten the name for easier reading) has contacted you to see if you can conduct a thorough examination of their site to make recommendations for improving conversion.

So where do you begin?

A. I would start making a Persona.

B. I would ask if they have any Personas.

C. I would review their Google Analytics data.

D. I'm not sure. I wasn't really paying attention to the overview chapter of the book.

This one is pretty easy, right? The first thing you should do is to establish if they already have Personas or not. If they do, then carefully review them to make sure they are Design Personas and have critical tasks associated with them. If not, you'll have to advise them that it will be necessary for you to do some research to make a Persona or Personas for this project.

Typically before I agree to conduct a UX optimization project I inquire if the firm has Personas. If they do, I ask that they be sent to me so I can examine them. I want to know before providing my proposal whether I'll be spending time creating a Persona or simply using one that was already created for UX design work.

In this case, they don't have UX Design Personas.

Knowing this, you decide to do research including contacting and interviewing older adults you know, reviewing related active ager fitness websites, and evaluating demographic and psychographic data the company provided you, as well as reviewing the same types of data from the website's Google Analytics reports.

You create your Persona based on the research you've done, which brings up the next question.

Your Persona should include which of the following elements?

A. Critical tasks the Persona must conduct to be successful

B. A story about why the Persona might be looking for this service

C. Contextual information like typical devices used, their domain expertise, and computer skills

D. Their specific age, household income, and geography

E. A, B, and C

F. All of the above

G. Wow, this is harder than I thought. Can I use my "phone a friend?"

This one is a bit more difficult to answer. To understand what information to include in a Persona, remember that you are looking for enough information to be able to say that the critical tasks of the Persona are representative of the critical tasks most users must do to be successful on the website. In this case, the answer is E: that A, B, and C should all be included in your Persona. D (specific age, household income, and geography) is not typically used in UX Design Personas.

The information you should include in your UX Design Persona includes

- Critical tasks associated with the Persona

- A story of the typical reason why the Persona would want this product or service

- Contextual information about devices used by the Persona

- The domain expertise and computer skills of the Persona

As an example, Figure 8-3 is the Persona I created for the Shop team. I reviewed this Persona with their team prior to conducting the analysis.

Persona - Jessica

Health-Focused Senior:

Jessica is a 64-year-old mom and grandmother who is focused on her health. She's looking for a way to keep moving, lose weight, and stay in shape. Her doctor has advised her to walk and use weights and she's also been told to modify her diet by reducing her salt. Jessica has not worked out in many years and is unfamiliar with senior-specific needs. Importantly, she's on a fixed income so price is important. She needs to find a health-focused solution that helps her get in shape.

Jessica – Health-Focused Senior

Education	H.S.
	College (partial)
	College
Job situation	**Not employed**
	Part time
	Full time
	Full time Student
Computer type	Smart Phone
	Tablet
	Desktop
Computer tools	Advanced applications
	Coding tools
	Email
	Web browsing
	Word processing
Computer skills	**Limited**
	Moderate
	Advanced
Domain expertise	**Low** Medium High

Jessica's critical tasks:

1. Find an exercise and diet health program.

2. Evaluate the cost and benefits of the program.

3. Buy the program that meets her needs & budget.

Figure 8-3. *The Jessica Persona created for the Shop.MyEverGreenWellness.com optimization project*

It is important to review the Persona you develop with your client prior to doing any evaluation so you and your client are clear on whom the primary user is and the critical tasks that need to be evaluated to optimize the conversion of the website.

Evaluating Behavioral UX Data

Now that you've created your Persona and your client has approved it, you can start examining the behavioral UX data you've already collected to look for patterns. Your goal is to identify the "what's happening" information. Which brings me to another question.

The behavioral UX information you should evaluate includes all the below with the exception of?

    A.   Screen resolution

    B.   Device overview

    C.   Usability testing results

    D.   Mobile device overview

    E.   Behavior flow

    F.   Session duration

    G.   Ummm, I wasn't actually paying attention. Sorry!

For all of you who were paying attention to the behavioral UX data chapter, you'll be able to identify that everything on the list belongs with the exception of C (usability testing results) and G (you were paying attention, right?).

All of the data listed is quantitative data, but usability testing results are qualitative data. The qualitative data (i.e., "why it's happening" data) will come later in your analysis.

Figure 8-4 is an example of one of the behavioral UX data elements examined during the analysis of the Shop project. This shows the screen resolution data for the site. Based on this data, it appears that iPads are the highest number of screen resolutions sizes, followed closely by Android devices.

Website Data: Screen Resolution

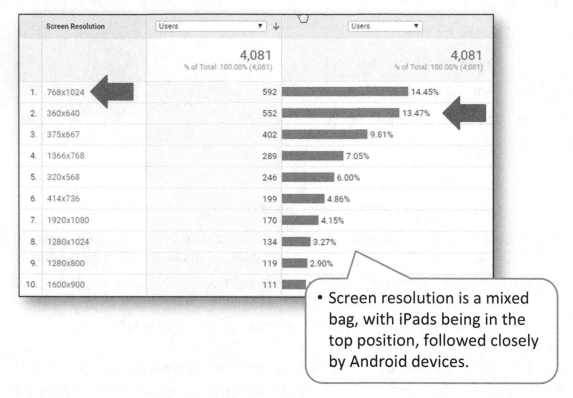

Figure 8-4. *Screen resolution listing iPad as the top resolution*

By evaluating this and the other common behavioral UX data elements, you can start to identify what devices and screen resolutions are used. This helps you understand the amount of screen content above and below the fold, among other things. This and the related behavioral UX data make it easier to see the patterns of usage on the website.

Let's continue your evaluation of the behavioral UX data.

Understanding where the page fold is for most website visitors is important because:

A. It's not important. The page fold is a myth.

B. I can understand where people leave the page.

C. I can use the data to identify what content should be moved lower down the page.

D.   I can identify "hidden" content that may not be viewed as often.

E.   Why, oh why, didn't I pay more attention to the behavioral UX
     data chapter?!

The answer is, by understanding where the page fold is for most website visitors, you can identify hidden content that may not be viewed as often. If that content happens to be important calls to action like a "Learn More" or "Buy Now" button, then you should note that information as being a potential issue to further evaluate. So the most correct answer in this case is D. You can identify any hidden content or calls to action that may be receiving fewer impressions and thus lower engagement.

Figure 8-5 provides a demonstration of the page fold for the most common visitors using the screen resolution and device information provided by the behavioral UX data. It becomes clear that some calls to action and other important elements are lower, in some cases much lower, on the page.

Website Data: Page Fold

Figure 8-5. *Displays the page fold for the most common website visitors*

Although it is true that scrolling occurs more now than it did 10 years ago or so (especially because of mobile devices), it is still a best practice to not force your visitors to rely on their memory of what content is where as they scroll up and down a page. This adds to their cognitive load, and anything that forces visitors to have to stop and think or try to remember where something was is not a good thing.

Placing the most important elements above the fold ensures that those items are seen by the maximum number of visitors.

Keeping important content above the fold helps reduce cognitive load. So understanding how the page fold is impacting important content can make a big difference when looking for issues causing poor conversion.

Other Behavioral UX Data

As part of your evaluation of the behavioral UX data for the Shop experience you'll also want to analyze many other data points.

What additional behavioral UX data elements should you evaluate?

A.   All available data, and sort out what's important later

B.   Top content, most and least visited pages, time on page

C.   Page flow including which pages cause pogo-sticking (back and forth page movement without progression) and bounces (landing on a page and then leaving the website)

D.   Time spent on key pages, scroll maps, click maps, and heat maps of page elements and especially calls to action that were or were not clicked

E.   All the above

F.   B, C, and D only

As enticing as it may be to just review all the available data and sort it out later, you probably don't have enough time to do that. And not all data is equally important. The correct answer is F.

After you've conducted several of these behavioral UX analysis reviews you'll realize there are certain pages in a website or app that need more focus first because they have a bigger impact on the overall user experience and conversion.

Evaluating the top content accessed, most and least visited pages, and time on page is helpful for determining a number of things including

- Are visitors finding the content they desire?

- What pages cause visitors to back up or leave?

- How long do visitors stay on each page? Are they consuming the content or scanning and moving on?

There are other important factors that should be evaluated as well, including

- How far down the page do visitors scroll?

- What elements on key pages are visitors clicking or not clicking?

- How many clicks are the primary calls to action receiving vs. the rest of the links on the page?

By the way, another version of the last bullet point (what is being clicked) is, What elements on a page are being clicked that are actually not even clickable links? That is very handy information to know when looking to improve task flow and conversion!

Because this is an eCommerce website case study, you have continued your exploration of all the above data (and more) and are now ready to look at the next set of data: eCommerce behavioral UX data.

eCommerce Behavioral UX Data

Because this is an eCommerce site, you will evaluate the behavioral UX data inside the buy-flow, including the shopping cart and each page within it.

What information should you review when evaluating the behavioral UX data of a shopping cart?

A.   Number of sessions for each page

B.   Number of abandonments for each page

C.   Number and type of links to external pages in the buy-flow

D.   Feedback form results obtained from inside the buy-flow

E.   A, B, and C

F.   I need more coffee to answer this question.

When looking at the behavior in the buy-flow, it's usually a good idea to determine the number of sessions for each page, the number of abandonments per page, the rate of abandonment per page (expressed as a percentage usually), and the number and type of links (if any) connecting to pages outside the buy flow. So answer E (A, B, and C) is the correct answer. The feedback form results (D), if there are any, are typically part of the qualitative assessment that comes after the quantitative analysis.

Figure 8-6 is a typical shopping cart abandonment rate chart. This is a visual representation of the number of sessions moving through each page of the buy-flow and cart. Abandonment rates (the percentages below the number sessions data) are useful for determining where abandonment rates are high.

Figure 8-6. *Abandonment data in a typical buy-flow*

Knowing that certain pages have high abandonment rates, like the Add to Cart page demonstrated in Figure 8-6, helps you zero in on the pages experiencing the highest issues in terms of people moving through tasks. Starting optimization efforts on those pages usually can improve eCommerce results quickly, and sometimes dramatically.

If the product or service the firm provides is focused on a select target demographic, as Shop is with older adults, it's a good idea to check if that demographic is actually coming to the site and spending time there.

Google Analytics provides a handy way to determine the gender and age of visitors. The Audience>Demographics>Overview Report in Google Analytics will identify the greatest number of visitors by age group and by gender. Figure 8-7 shows the demographic data for Shop.

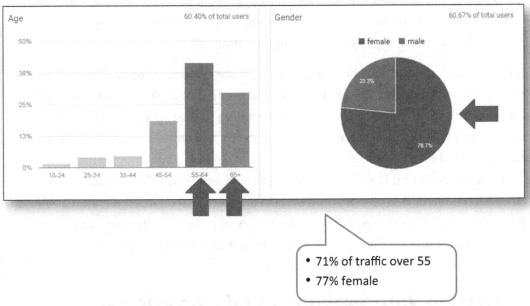

Figure 8-7. *Age and gender of the most common visitors to Shop*

This data is helpful for determining the age and gender of your Persona, as well as validating who is visiting the website. The demographic data is important for you to consider when evaluating the user experience of the site for several reasons. Knowing that the site is focused on providing health improvement courses to older age groups (especially older adults) and knowing that mostly women visit the site has several ramifications to your research.

Why is the age and gender of website visitors important to your Shop UX research? Choose the BEST TWO ANSWERS.

A. This data explain why visitors are or are not buying the products.

B. The data validates the Personas.

C. I can use this data to focus on specific aspects of the user experience that are impacted by age and gender.

D. Any differences in the target audience versus website visitors explain why conversion is low.

E. Women are better shoppers than men, so more women visiting a website should always indicate higher conversion.

This is a difficult question because there are a few answers that are partially right. The key thing to remember is you are in the quantitative phase of testing, the "what's happening" phase. So although it may be compelling to draw conclusions as to why visitors are not buying, or why conversion is low, you actually do not have enough "why it's happening" data to draw those conclusions.

Thus, answers A and D are not as correct as answers B and C. Oh, and by the way, if you included answer E as one of your two best answers, I'm a tad bit worried about you. Women are certainly good shoppers (as are plenty of men) but drawing conclusions about conversion based only on the gender of a website visitor is pretty dubious.

Summarize Your Personas and Behavioral UX Data Findings

So we don't end up with this book having more pages than *War and Peace* I'll skip ahead in the case study!

You have now reviewed enough data that you can see some patterns starting to emerge. These patterns include

- Understanding who the typical users are (Personas)

- Pages that experience low or abnormal activity

- Task flows that have issues in certain places

- Interactions with lower activity than expected

- Mismatches between target audience and visitors

After completing your evaluation of all the behavioral data including the above case study examples and more, you should summarize your findings into short, bit-sized pieces of patterns.

The goal of your summary of the behavioral UX analysis is to help your client or team understand the "so what" of the "what's happening" data, and where you are finding these patterns.

You should include

- Who you are evaluating (the Persona or Personas)

- Irregular behavior flow patterns that hint at pages to evaluate further using usability data

- Other patterns of usage that indicate where users are experiencing difficulty or abandoning a critical task

- Your next steps (especially around how you use this data to identify what qualitative data to analyze)

For our friends at Shop.MyEvergreenWellness.com you might summarize the findings from your detailed investigation of the behavioral UX data as demonstrated in Figure 8-8.

Behavioral UX Data Findings

1. Women (55+) are visiting on mobile devices and iPads but are leaving. A small minority are coming back to purchase via desktop.

2. Behavior flow indicates typical path is Little Black Dress > Cart > Little Black Dress, which is often associated with a lack of key information and/or trust on the Little Black Dress page.

3. Data suggests visitors are exploring the site using mobile devices but are purchasing via desktop or tablet, which hints at visibility issues (i.e., needing larger screens to conduct purchases).

4. Based on the above, we will focus the analysis on optimizing the experience for Persona Jessica and her use of a tablet (second-most-used screen resolution, second-most-purchased device type).

Figure 8-8. *Behavioral UX data findings summarize the big picture and identify next steps*

You'll more than likely have one or two more summary slides. One slide may perhaps be focused on the eCommerce issues found and another on any additional patterns of unusual behavior that should be investigated.

A common question I'm asked at this point is

> *"Do I show the team the findings so far? Or do I wait for the additional information coming from the usability testing and other qualitative data?"*

I believe you already know the answer to that question, assuming you've read the prior chapters!

The answer is, it's best to wait to present your behavioral UX data analysis because so far all you've uncovered is the "what's happening" quantitative data. You still don't know "why it's happening" qualitative data.

For that, you need to take your quantitative patterns and findings from the behavioral UX data and use that information to target your attention on specific qualitative usability testing data.

Your goal is to use the "what's happening" data to help you learn "why it's happening." And that is the next chapter! So grab your coffee, and I'll meet you over there!

CHAPTER 9

Putting It All Together: Usability Testing Data Analysis and Recommendations

With the quantitative behavioral UX data analysis complete, you have identified patterns that indicate "what's happening." You will use the "what's happening" quantitative analysis to help you determine on which pages and with what tools you should evaluate the "why it's happening" qualitative data.

Let's continue with the case study of the Shop.MyEvergreenWellness.com eCommerce website to see how usability testing and related qualitative data will help you identify the WHY.

Usability Testing and Qualitative Data

Where you begin your analysis with usability testing and related qualitative data will in great part depend on what patterns you found in the behavioral UX data analysis.

For the team at Shop, because this is a relatively new brand and because the patterns of usage seem to indicate higher abandonment in and out of the funnel, a 5-Second Test is in order.

© W. Craig Tomlin 2018
W. C. Tomlin, *UX Optimization*, https://doi.org/10.1007/978-1-4842-3867-7_9

I feel the need, the need for another pop quiz!

Why use a 5-Second Test? You use a 5-Second Test to evaluate how well the website (choose all that apply):

    A.   Loads within 5 seconds

    B.   Communicates who the brand is

    C.   Helps visitors understand what products or services are provided

    D.   Provides visitors with an understanding of what's in it for them

    E.   All of the above

The answer is a 5-Second Test is extremely helpful for evaluating three important items: do visitors remember the brand, do they understand what products and services are offered, and perhaps most importantly, do they understand how the products or services offered may benefit them and/or solve their needs (i.e., what's in it for them)? For this quiz, if you picked B, C, and D, you are my star pupil and deserve a gold star. Option A (loads within 5 seconds) is not what is evaluated with a 5-Second Test.

Without those three critical pieces of information being relayed to visitors, the website will experience reduced trust, higher abandonment rates, and lower conversion.

For the case study with the Shop website, you have decided to test the home page using the 5-Second Test. Because this site is focused on providing health improvement specifically for older people, it may be important to know whether the visitors realize that the products are focused on their age group. If they are unaware that the training program was specifically created to benefit them, they may be less inclined to buy it.

If that's the case, this could be one of the key reasons why you see the higher abandonment rates as reported in the behavioral UX data.

So given the above, you want to use the 5-Second Test to learn how well the website communicates

- What product or service is provided

- The targeted age group of the product or service

- The name of the company

Figure 9-1 provides a brief overview for the setup and administration of the 5-Second Test. I typically provide this simple overview of the 5-Second Test methodology as part of the final analysis presentation I provide the client. This overview helps the client understand what the 5-Second Test is, how it works, and why I am using it for their website analysis.

5-Second Test

- The LP was tested using the 5-Second Test. Testers were shown the page for 5 seconds.
- After 5 seconds, the image was removed and they were asked
  - Q1: What product or service do you think this firm provides?
  - Q2: What age group is this product or service for?
  - Q3: What's the name of the company associated with this website?
- Visitors were recruited at random and were not aware of the firm or its products prior to testing.

Figure 9-1. *A brief overview of the methodology used for the 5-Second Test*

To conduct the 5-Second Test you'll use the same methodology you learned in prior chapters. You want to evaluate the home page of the site for the three questions.

After you have set up and run the 5-Second Test you can now provide the analysis for the results. Pie or bar charts are very effective for providing the data and analysis of the results. Figure 9-2 shows the results for the first question from the Shop home page.

5-Second Test Results

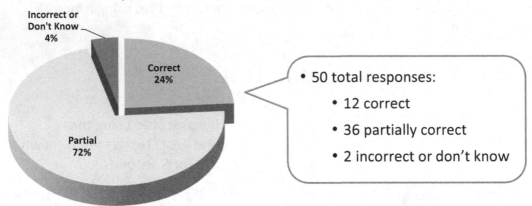

Q1: What product or service do you think this firm provides?

Figure 9-2. *The 5-Second Test results for the product and service question*

For the next 5-Second Test question regarding what product or service is offered (displayed in Figure 9-2) the results of the 5-Second Test reveal that about 24% of the participants were able to correctly identify the product for this website (health improvement programs for active agers).

Unfortunately, the vast majority (72%) of respondents were only partially correct. They reported that they thought the product had something to do with health or seniors, but they did not understand it offered healthy lifestyle content and fitness programs for active agers.

The good news is only 4% were unable to correctly determine the product or service. **So what do you think the results tell us?**

A.   People are just plain clueless.

B.   The website may not be communicating what product or service is offered as well as it could be.

C.   There's not enough data yet to determine anything.

D.   If only 4% of people don't get it, that's actually good and the website is communicating effectively.

For all of you who guessed answer A, please don't give up on humanity just yet! For the rest of you, the results of this question actually reveal that the website is somewhat effective at communicating what products and services are available, but could be doing better.

If only one quarter of website visitors clearly understand the product or service in 5 seconds, that is a strong signal the website could be communicating better. This is an indicator that improvements in how the product or service is communicated can be made and should have an impact on conversion. So in this case, answer B is the correct answer.

The second question in the 5-Second Test focuses on the age group. Do visitors understand what age group the product or service is for?

For some products, it does not really matter if the website visitor understand what age the product or service best benefits, often because any age group can use the product to solve their needs.

But for this program, which is specifically targeting seniors who want to improve their health, it's important for visitors to understand this information. After all, it's rather unlikely someone would buy a product if they don't know who that product is for.

Figure 9-3 provides the results of the second question from the 5-Second Test. The vast majority of respondents assumed the product or service was for adults, but only a handful (six out of 49) correctly identified that this product and service is for older people and seniors.

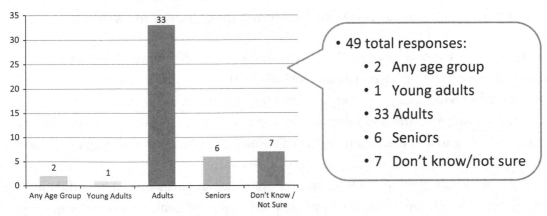

5-Second Test Results

Q2: What age group is this product or service for?

- 49 total responses:
  - 2 Any age group
  - 1 Young adults
  - 33 Adults
  - 6 Seniors
  - 7 Don't know/not sure

Figure 9-3. *Answers to the age question in the 5-Second Test reveal that respondents overwhelmingly thought the product was for adults, not seniors*

Based on the results as displayed in Figure 9-3 you can assume most visitors are not clear about the targeted audience. This could be one of the reasons why you see higher abandonment and lower purchase rates than expected. This data also strongly suggests there is an opportunity for the website to better communicate what age group the product or service is targeted for.

Finally, the third question in the 5-Second Test was the brand question. Figure 9-4 shows the results for the question about the name of the company associated with the website.

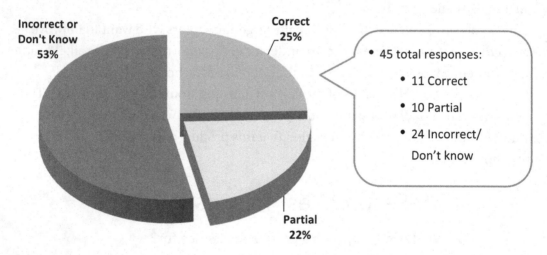

Figure 9-4. *The 5-Second Test results for brand awareness and retention*

Among the respondents, only 25% correctly identified the brand. Only 22% partially got the brand correct, using names that included either "Evergreen" or "Wellness." But the majority of participants (53%) completely failed to recall the brand.

This data suggests that branding can be improved on the website so visitors are more aware of the brand and can retain the brand name for hopefully much longer than just a few seconds.

If the brand is not memorable, even after only 5 seconds, it would seem unlikely visitors might remember it should they desire to investigate healthy lifestyle programs in the near future. In addition, a brand that is not memorable might cause trust issues, which could explain why some visitors abandon the process of purchasing a product.

So now that you have the results from your 5-Second Test, how would you summarize these findings? What does this data tell you to use to help optimize the website?

What are your recommendations based on the evaluation?

A. Use taglines and body copy as appropriate to clearly define the product and how it helps seniors get healthier.

B. Add additional senior-oriented visuals and copy in the content to clearly identify that the primary audience for these products is seniors.

C. Add branding elements and enlarge the brand above the fold and use the brand in copy/visuals where applicable to reinforce the brand.

D. Use a burning, spinning logo. That will get their attention!

E. A, B, and C

When faced with 5-Second Test results that indicate low product or service understanding, low comprehension of who the product is for, and difficulty remembering the brand, you can suggest several optimizations:

First, to help visitors understand the product or service, you need to use taglines and other one- or two-sentence statements high up on the page (the global header is a good choice) that clearly defines WHAT is being offered and HOW it helps the visitors. Taglines may sound like marketing fluff, but in reality they are the workhorse of the product and service awareness cause. Sprinkling taglines or simple bullet-point statements of what is offered throughout the design treatment as appropriate should help improve results.

Second, to improve visitor comprehension of whom the product or service is for, it's important to include that information in taglines, which are short statements in the copy and in visuals. There are many examples of websites that use stock photography images of random smiling people who may be doing something that might or might not relate to the product or service. Those images are often poor at communicating who the product is for. Any image chosen for a web page should be carefully thought out to make sure it clearly communicates who should be using the product or service and how it helps them.

Third, for brand recognition and retention, the logo, tagline, and other short statements reinforcing the name of the brand in the copy (where appropriate) are crucial. Logo placement, tagline placement, and use of the brand in copy will greatly impact the ability for your visitors to remember the brand. And a memorable brand is more likely to garner repeat visits and potentially be remembered when it is time for the visitor to purchase the goods or services offered by that brand.

So the answer to the question of what recommendations to provide is E (which is A, B, and C).

Using taglines to help define who the product is for and how it helps them is a must for communicating what the product or service is.

Carefully choosing visuals and adding copy in the content to clearly identify whom the product or service is for and how it helps them is also critical.

Finally, adding branding elements or enlarging the brand above the fold and where applicable in the body copy is an important part of helping to reinforce brand awareness and retention.

The Power of 5 Seconds

Over the years I have come to feel that the 5-Second Test and subsequent analysis and optimization based on the results is a very powerful tool in the conversion optimization toolkit. Most, if not all, of my clients have agreed with me. Yet the 5-Second Test is one of those tests that most UX practitioners either don't know about or don't use often.

The beauty of a 5-Second Test is you can conduct the test very quickly and repeat the test again and again as you make optimizations to your website. This enables you to quickly acquire and use data to continually improve results.

So the next time you are conducting UX research and have questions like those we just covered, I sincerely hope, no, I fully expect you will use the 5-Second Test, just like I do!

Now that I've addressed some of the more critical elements that make up the design of the page with the 5-Second Test, let's look a bit closer at the text and calls to action (CTAs) on the page.

Body Copy and CTAs

Figure 9-5 is a close-up of a Shop page as seen on an iPad. What elements on the page may help or hinder a senior who is interested in the product or service?

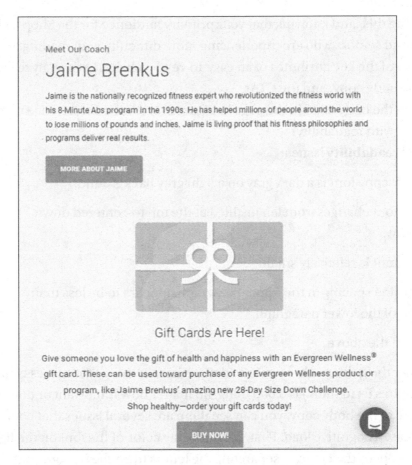

Figure 9-5. *Close-up of the body copy of a page on the Shop website*

You may have heard the phrase "the devil is in the details" and for websites this is especially true. From the "what's happening" data you obtained earlier, you know there is higher abandonment rate than desired, and many pages in the site are experiencing lower engagement than anticipated.

I often like to say

"Engagement comes with comprehension, and comprehension comes with readability."

If the content is not readable, due to a variety of reasons, then comprehension will be limited and thus engagement will suffer.

Put another way,

"If they can't read it, they won't understand it. And if they can't understand it, they won't buy it!"

So knowing this, and knowing that your primary audience for the Shop website is older adults and seniors who are experiencing more difficulty with reading, let's have a look at some of the key attributes of an easy-to-read website. Let's analyze the fonts, especially the body copy, and the CTAs.

Looking at the body copy in the page in Figure 9-5, what do you notice that may be causing issues with readability?

Potential readability issues:

A.   Body copy font is a dark gray on a light gray background.

B.   The font changes from left justified at the top to centered down below.

C.   The font is relatively small.

D.   The line spacing in the upper paragraph appears to be less than that of the lower paragraph.

E.   All of the above

Remember the golden rule of website design that anything that causes extra cognitive load (i.e., requires us to think) is a bad thing, no matter how important or how minor.

In the case of the body copy, you can see there are several issues that could be causing increased cognitive load. First, the dark gray color of the font on the light gray background reduces the contrast separating the letters from the background. That increases cognitive load.

Likewise, switching the font from left justified at the top to centered below also causes cognitive load.

And just a point about centered text: If the location of the beginning and ends of each line are not aligned along the left and right sides of the copy, the ragged way each line breaks (some lines longer, some shorter) will cause some additional cognitive load. Our eyes are forced to search for the beginning of each line, after having finished with the line above. That is why cognitive load increases for centered text that is not justified on both the left and right sides of the lines.

For seniors who already have difficulty with vision and most likely need reading glasses, the font may well be too small.

Finally, there appears to be differences in the spacing between each line for the body copy paragraph at the top of the page vs. that at the bottom of the page. That is another cause of increased cognitive load.

Taken all together, each of those issues can cause increases in cognitive load, making the page harder to read than necessary. So the correct answer is E (all of the above).

By improving the readability of the body copy (and the rest of the text on the page) you can reduce cognitive load and help visitors read and understand the information. This can help to increase comprehension and potentially engagement and ultimately sales.

The same issues for causing extra cognitive load to body copy apply equally to CTAs. In this case, the CTAs are the same color and are positioned left justified as well as centered on the same page.

The general rule of thumb for CTA buttons is the most important CTA (for eCommerce it's usually the "Buy Now" button) should always be a unique color vs. any other CTAs. Likewise, the positioning of the CTA should align with the page such that it is very easy to see, obvious, and in the same place whether you are above the fold, in the middle of the page, or at the bottom.

In Figure 9-5, the CTAs for "More About Jaime" and "Buy Now" are the same color and in different positions on the page. Since you can assume the "Buy Now" button is far more important to the Shop team, the recommendation is to make sure it is a unique color and also the most prominent button on the page.

Figure 9-6 shows the resulting analysis slide presented to the Shop team with the recommendations for font and CTA optimizations.

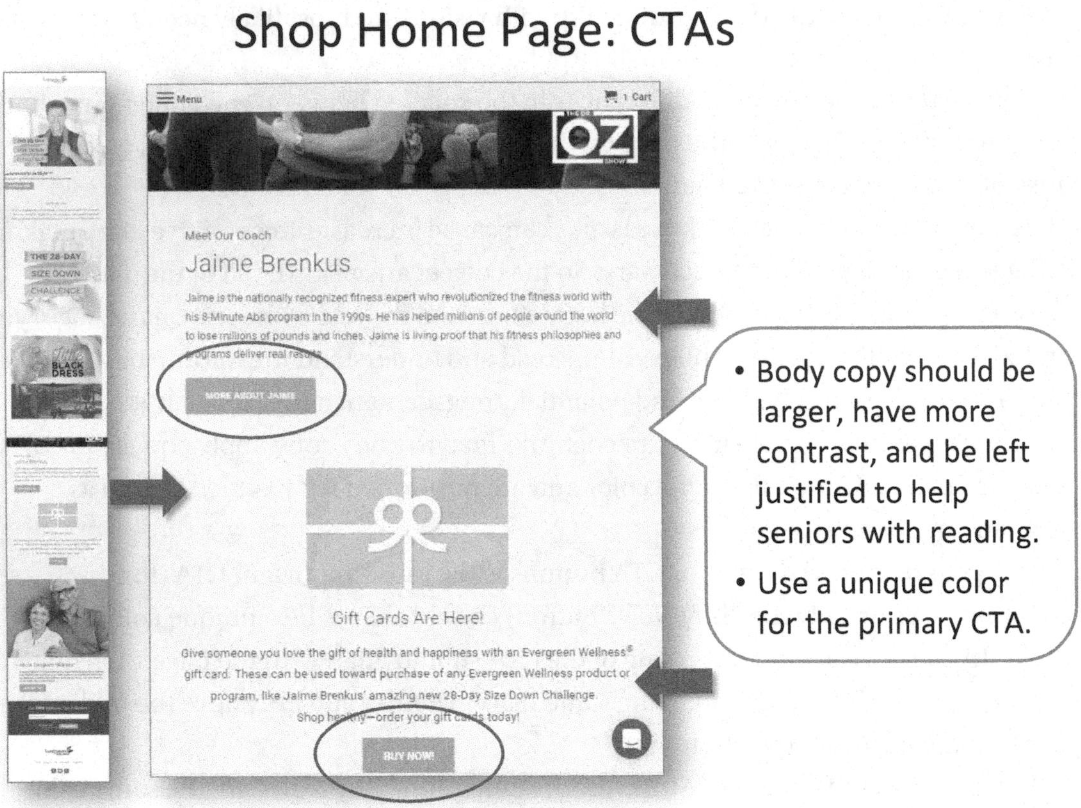

Shop Home Page: CTAs

Figure 9-6. Analysis slide with recommendations for body copy and CTA optimizations

The Eyes Have It: Automated Eye Tracking

Another handy tool to evaluate the WHY of the behavioral UX data you analyzed is eye tracking. Eye tracking tools capture fixations, those momentary pauses of the eye as you look at objects. By analyzing the number of times fixations happen on various elements on a webpage you can obtain heatmaps of activity. Areas with more fixations are colored in reds, representing higher fixation areas. Areas with fewer fixations are colored in yellows to greens to blues as the number of fixations drop. Areas with no fixations have no color.

For your Shop website evaluation, you have already identified behavioral UX information that suggests visitors are bouncing back and forth between pages. You also know most visitors don't clearly understand the product or its targeted audience.

Let's try to understand the why of this behavior based on what objects on the page are or are not attracting attention.

In a perfect world, I would use real eye tracking studies to identify potential areas on the page that are capturing attention. But it's not a perfect world, and eye tracking equipment and studies are relatively expensive. They also take a fair amount of time to set up, run, and analyze.

For situations in which I don't have the money and/or time, I like to use automated eye tracking. One of the automated eye tracking tools I like to use is Feng-Gui.

It's important to note that automated eye tracking is not the same as conducting real eye tracking studies with actual people observing a website. The automated eye tracking results are based on algorithms that take into account many factors of what attracts attention. For example, we are all wired to automatically look at a human face. Likewise, if the eyes of the person whose face we are observing are looking in a direction, we will automatically look in that direction too. There are other factors involved in how the algorithm works, but that is outside of scope for this book, so we'll move on.

The key point to remember is automated eye tracking data is more of a suggestion as to what may be capturing attention than actual fact. I always make sure I let my clients know this, and so should you if you decide to use this tool in your analysis.

Figure 9-7 is one of the pages on the Shop website prior to the automated eye tracking test. Note the face and the brightly colored Chat icon on the dark background in the lower right corner of the image.

Figure 9-7. *Shop page prior to automated eye tracking testing*

Figure 9-8 is an example of the same Shop page with the results of the automated eye tracking displayed as an overlay heatmap. Note the higher fixation area on the face (as we would expect). And also note the higher fixation area on the Chat icon on the lower right caused by the high contrast of a bright color on a dark background (also as we would expect). The light areas or clear areas in the heatmap, especially those on the text below the video, may indicate the face on the video and the Chat icon might be pulling attention away from the heading and body copy.

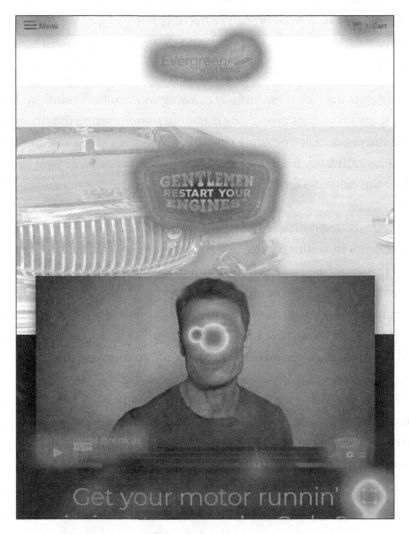

Figure 9-8. *Automated eye tracking results for a Shop page*

What key findings might you draw from the eye tracking data?

A.   Don't use faces on websites; they attract too much attention.

B.   The face may be pulling attention away from the logo (brand).

C.   The face may be pulling attention away from the headline and body copy.

D.   There is not enough data to suggest any findings.

    E.   Help me, Obi Wan Kenobi. You're my only hope in getting this
        question right.

    F.   B and C

The data suggests the face is potentially drawing most of the attention. Knowing
that prior data suggests visitors are having difficulty remembering the brand and
understanding the product, it's possible that the face in the video thumbnail above the
fold is pulling attention away from the logo at the top of the page, as well as the headline
and copy below the video. So other than using the Force, the correct answer for this quiz
is F (B and C).

The data suggests that perhaps using a larger logo may help the audience better
remember the brand. In addition, clearly stating the purpose of the site above the video,
or better yet in the thumbnail image for the video itself, may help communicate the
purpose of the site as well as the brand.

In addition to the Shop page above, all of the other primary pages for the website
were evaluated with automated eye tracking. Figure 9-9 is an image of the slide from
the Shop analysis presentation provided to the client highlighting the results of the
automated eye tracking for the Programs page of the website.

The heatmap in Figure 9-9 and the rest of the heatmaps from the other pages of the
website have similar results. Most of the attention is potentially being drawn to areas
that do not clearly identify the brand or the purpose of the site and products. This eye
tracking pattern is consistent across all of the pages.

Programs Page: Eyetracking

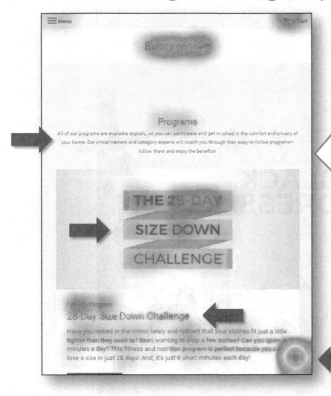

- Eyetracking heatmap implies small centered text is harder to read.
- Image may be attracting attention, heading attracting attention (all good).
- Strong chat icon may be pulling attention due to strong color/shape.

Figure 9-9. *Analysis of automated eye tracking on the Programs page*

One final point about eye tracking studies is the data can be displayed in a variety of ways. Sometimes displaying the data using a different view can be more effective than displaying the typical heatmap. An example of this is the opacity view.

Figure 9-10 is a view of the top of another page of the Shop website, the Little Black Dress page. This page promotes a weight loss exercise and diet program for senior women.

Figure 9-10. *The Little Black Dress page*

Using an eye tracking opacity view can help visualize what objects on the page *should* be attracting attention but may not be. This can sometimes be a better way to display data, as it may help point out where design elements may not be working as anticipated.

Figure 9-11 is the opacity view of the automated eye tracking results for the Little Black Dress page. Note that this is a different view of the data compared to a heatmap view. In the opacity view, the elements that potentially are attracting attention are viewable, much like viewing a scene through a window covered with black paint. Areas in the window where the paint has been wiped away (to see "through" the window) represent areas of attention. Areas that are opaque and cannot be seen are areas of little or no attention.

Figure 9-11. *Opacity view of the automated eye tracking data for the little black dress page*

By evaluating the results of the automated eye tracking data in opacity view for the Little Black Dress page, you can observe some interesting patterns.

Potentially the words "little black dress" are attracting attention. But missing may be additional clues about the product, such as that the product will help a person fit into a smaller size dress in two weeks. You may even wonder if potentially some visitors are confused as to whether they might be able to buy a little black dress on the website, or if there are other reasons for the little black dress text on the page.

Again, anything that causes unnecessary cognitive load is a bad thing when it comes to website design. The data from the opacity view would seem to suggest improvements could be made to the video thumbnail to reinforce that this is a fitness and nutrition program focused on helping women look and feel better in their clothes.

Using data gathered from automated eye tracking studies, coupled with the other usability data you have already analyzed, you can start to see patterns that may explain the WHY of the behavioral UX data you have already evaluated. This "why it's happening" data helps you focus on key elements of the user experience and design that may be causing lower conversion. You can also formulate theories for what to do to improve the site that are based on this data. This is what makes this methodology so precise, and so powerful.

But, as any good infomercial will say,

"But wait! There's more!"

Pushing the eCommerce Shopping Cart (to Better Performance)

In addition to the informational website pages you have analyzed, eCommerce sites have additional pages to review. For eCommerce sites, it is very important to also evaluate the shopping cart and checkout process. This makes sense because very small increases in conversion in the shopping cart and checkout process can have dramatic increases in revenue.

As a general rule, anything that distracts people from going through the shopping cart process is bad. Anything that allows people to leave a shopping cart is considered bad. Anything that causes additional cognitive load in the shopping cart is bad. Anything that distracts or pulls attention away from the task of completing the shopping cart purchase process is bad. Put another way,

Anything in the shopping cart and checkout that is not necessary in helping people complete the purchase is bad.

As with all generalities, there are a few exceptions to this rule. I'll get to those in a moment.

Figure 9-12 is a screenshot of the shopping cart for the Shop website. Think about what you know about keeping all unnecessary distractions out of the shopping cart.

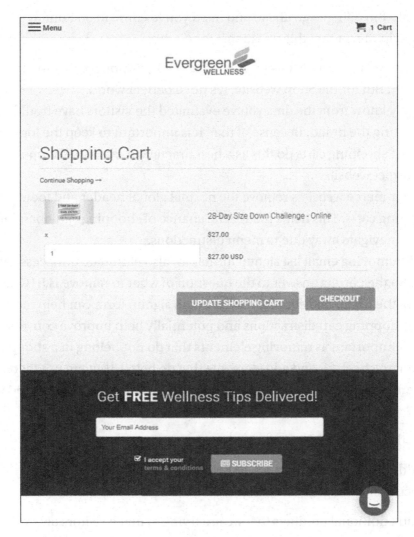

Figure 9-12. *Shopping cart page of the Shop website*

What elements, if any, would you identify as potentially being distractions that might impact conversion?

A.   Nothing. It all belongs in the shopping cart.

B.   The logo

C.   The menu

D. The email sign-up form

E. C and D

F. Yeah, I'm going to go ahead and need you to come in on Sunday
    to help me answer this question.

In a perfect world, the only design elements in the shopping cart support the process of checking out. But for the Shop website, it's not a perfect world.

You already know from the data you've evaluated the visitors have trouble identifying and remembering the brand. Because of that, it is important to keep the logo and branding. Most shopping carts do this as a best practice to reassure shoppers they are still on the correct website.

Often eCommerce websites remove the normal global header and footer navigation in their shopping carts. This helps reduce the chance of shopping cart abandonment by having people navigate away due to menu distractions.

Likewise, removing email list signup forms can also eliminate unnecessary shopping distractions. So the correct answer to the question of what to remove is E (C and D).

Removing the menu and eliminating the email signup form can help reduce unnecessary shopping cart distractions and potentially help improve conversion.

But just as important as removing elements that do not belong in a shopping cart, it's also important to check for and add elements that do belong but are not there.

What elements, if any, would you suggest adding to the page displayed in Figure 9-12?

A. Nothing, keep all distractions to a minimum

B. Add social proof icons, such as Better Business Bureau (BBB),
    secure site badges, star ratings

C. Add additional products that we are trying to make visitors aware
    of

D. Add links to the About Us page so visitors can learn who we are
    and thus trust us more

E. B and C

F. Why, oh why, didn't I take the blue pill and end all these
    questions?

Trust is a critical element of conversion, and this is even more true for eCommerce websites. If people on the shopping cart page of an eCommerce website do not trust the page, or the brand, then fewer people will purchase, so the number of conversions will be lower (all other things being equal).

Social proof is the tool most often used to help generate trust on a website.

Social proof can take many forms. Some of the most common forms include

- Testimonials from happy customers

- Star ratings from customer rating sites

- Icons of trusted vendors like the Better Business Bureau and others

- Security badges that display a confirmation that the information entered on a page is encrypted and safe from interception

Social proof is very important in the shopping cart and buy-flow of an eCommerce site. Without the aid of social proof, shoppers may have more difficulty trusting the site enough to give their credit card information as part of the checkout process.

So what's missing in the shopping cart? In Figure 9-12 you can see that social proof is missing and should be added. Including social proof icons from firms such as the BBB as well as the verification of the safety and security of financial information via security icons from third-party services are two easy ways to add social proof (assuming both are available for the firm). So the correct answer to the question is answer B.

So why not add additional products as recommendations to the shopping cart? And why not add links to the About Us page so people can learn more about the company? The answer is, anything that can cause a shopper in the shopping cart to become distracted and move out of the cart to some other page is potentially a bad thing.

Some eCommerce websites (including Amazon) go against this rule and offer recommended additional products in the shopping cart. That's worth A/B testing, but the assumption is you have enough data about the user to be able to fairly accurately suggest products that closely align with their needs and interests. That's actually a tall order for many firms, even with the improvements in AI and real-time use of big data for analysis.

My recommendation is to keep it simple, especially for a start-up with only a few products, and limit the shopping cart to only what is in the shopping cart.

Keeping distractions to a minimum in the shopping cart is always worth testing and optimizing. The resulting increases in conversion are almost always worth it.

There are other elements to evaluate in the buy-flow, including which pages and form fields are being abandoned at higher rates. Using this data, it's easy to determine which part of the buy-flow is experiencing abandonment.

By using either remote or in-person usability testing sessions to evaluate those specific pages and form fields, you can quickly identify why those pages or form fields may be causing higher abandonment rates.

I have also found that reviewing live chat transcripts in the buy-flow is another good way to determine the potential "why it's happening" data of abandonment. Live chat transcripts, assuming they are available in the buy-flow, will reveal what questions or concerns customers have when they are attempting to complete the purchase process. This can be a very helpful source of qualitative data useful for analysis.

Finally, a question I receive from time to time is on pricing and how much cause/effect it has on shopping cart abandonment. Some firms, especially those that can be flexible with pricing (services firms, for example), may not wish to divulge their pricing information on product pages. They sometimes attempt to only reveal pricing in the shopping cart or buy-flow. Let's explore this in a bit more detail.

The Price Is (or Isn't) Right

When asked where to place pricing information, or whether to offer pricing or shipping discounts, the definitive answer is

It depends.

No two companies are the same, so the issue of pricing, displaying prices, shipping costs, and related pricing information are dependent on many factors.

However, in general, there are several good rule-of-thumb guidelines that can be helpful when you evaluate the usability of a shopping cart and buy-flow in terms of providing pricing information. They include

- Is it clear how the product addresses the Persona's needs?

- Can the visitor evaluate costs vs. benefits easily?

- When pricing is displayed, does it include the most to least expensive costs in a clear way?

Figure 9-13 shows some of the pricing recommendations made for the Shop website. It is important to help a shopper understand which product may help them solve their needs in a very easy to understand way. Remember another key UX golden rule:

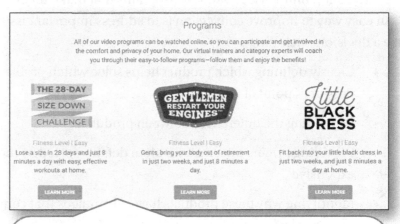

Figure 9-13. *Optimization recommendations for helping shoppers make a choice*

When offered too many choices, sometimes the easiest choice is to make no choice.
Helping shoppers understand the differences in products that seem similar is crucial to helping those shoppers make a choice and eventually become buyers.

For the Shop products, for example, what's the difference in the 28 Day Size Down Challenge and the Little Black Dress program? Both seem to be focused on helping lose weight to fit into a smaller dress size. What differences are there?

And what about the prices? Are there price differences between the three products? Studies have shown that displaying pricing sorted from most expensive to least expensive is often a good way to set cost expectations, which can often lead to higher order value, all other things being equal.

171

When evaluating products, shoppers also want to understand what it is about these particular products or services that make them unique or special compared to the competition. For a category like online fitness videos, there is a wide range of competitors and price points. This includes free videos on video sharing sites like YouTube to paid online videos through well-known fitness gurus. So what is it about these three Shop online video exercise programs that make them special vs. all the rest?

An easy way to improve conversion is to address important issues that help shoppers make a decision, including

- Clearly defining which product helps solve which problem (from the Persona's point of view)

- Explaining the differences between products

- Providing pricing information, often default-sorted by most to least expensive

- Identifying why these products have advantages that competitor products lack

I have found that probing the above by asking follow-up questions with participants during usability testing sessions is a good way to uncover the qualitative WHY of purchase consideration decisions either being made or not.

Useful Usability Testing

I saved the best for last, which is it's always a good idea to conduct either in-person or remote usability testing sessions of critical tasks with participants who match the Personas. Usability testing is one of the most useful tools, if not THE most useful tool, of qualitative data analysis.

As you learned from prior chapters, usability testing is the best way to bring data in from actual users to help identify task-flow issues. Nothing beats watching several users stumble over the same point in a task flow to help pinpoint exactly where a usability issue is causing poor performance.

Very often I am able to use the results of usability testing to confirm other qualitative data. This helps improve the results of your recommendations because if it looks like a pig, smells like a pig, and acts like a pig, it's probably a pig.

An extremely powerful and useful tool with usability testing is creating highlight reels that call attention to a key part of the process flow that is causing confusion, errors, or abandonment of a critical task.

I like to show three or so participants stumbling over the same issue to drive the point home. The big benefit of highlight reels is they don't take a lot of time to watch (try to keep them under 1 minute). The other key benefit is highlight reels clearly show the users as they stumble or commit errors, and nothing communicates better with internal audiences than videos of actual users having problems on the site.

So be sure to use the very useful usability testing tool when evaluating task flows on a website or app.

The ABCs of A/B Testing

Now that you have gathered your data and conducted your analysis of the behavioral UX and usability testing data, you are ready to present your findings to your client.

But there's one important element that needs to be included in your results presentation to your client, and that is

A/B test the recommendations!

Even though there is plenty of data to suggest that your recommendations will help improve optimization, there is no guarantee they absolutely will. The best way to test your recommendations is to conduct A/B testing of the recommended changes, and that's the advice you need to share with your client.

What is A/B testing?

A/B testing is a way of comparing two versions of a website page against each other to determine which one has better performance. A/B testing works by showing two variations of a page to site visitors at random. After enough data has been collected to provide statistical significance, analysis reveals which of the two variations performed better for a specific goal.

The control or A version is the original page with no changes. The B version is almost identical to the A version but has a single element that is modified. It's the modification of the original element on the page that is being tested.

The secret to A/B testing is to only change ONE element on the page. The reason why is if you try to test two or more changes on a page, you cannot be certain which change resulted in the difference in conversion between the two pages.

I always recommend the client A/B test any of the recommended optimizations I have made to ensure they improve conversion. I highly recommend you do the same.

As I like to say,

"A/B testing takes the assumptions out of any recommended changes because we all know to 'assume' is to make an ass out of you and me."

Conclusion: Usability Testing Data Analysis and Recommendations

Whew! You did a GREAT job covering a lot of material on usability testing data analysis and recommendations in this case study. Give yourself a pat on the back and have a beer, wine, or any of your other favorite beverages!

Together in this chapter we covered several important and useful tools and techniques for evaluating the qualitative "why it's happening" data. You knew what qualitative data to focus on based on the results of the quantitative "what's happening" data.

You've explored

- The **5-Second Test** and why it helps identify how well the site communicates the three critical elements of who you are, what you do, and why they should care

- The importance of **body copy**, **readability**, and **calls to action** to communicate effectively with your website visitors and help them engage with the site

- **Automated eye tracking** and how it can help you identify potential places in a design that are either not attracting enough attention or attracting too much attention

- An understanding of the eCommerce shopping cart and the pages and/or fields that may be causing higher abandonment rates including recommendations for optimizations based on **social proof** and best practices

- **Pricing** and how product descriptions, benefits, and unique attributes compared to the competition can help stimulate purchase consideration

- **Usability testing** and how highlight reels demonstrating critical task flow errors are very helpful for communicating where issues are occurring

- **A/B testing** and why it is very important to recommend using it to test and validate your optimization recommendations

Using the analysis techniques and tools mentioned in this and the prior chapter will greatly help you tell the story of both "what's happening" and "why it's happening." Using this data and the analysis you performed will provide more accurate and more detailed recommendations for optimizations.

How Do You Get to Carnegie Hall?

The old joke of "How do you get to Carnegie Hall?" "Practice!" is very useful at this point. Now that you've learned what to do, and how to do it using your use case, you now have to practice, and practice, and practice!

The more time you spend conducting behavioral UX and usability research and analysis, the easier it will become. And I think that once you've got the hang of it, you'll feel right at home conducting evaluations, analysis, and making recommendations.

In the next chapter, I'll sum it all up and leave you with a few observations, tips, and some of my strategies for conducting your analysis and optimizations. But for now, enjoy that beverage and I'll catch you on the flip side!

CHAPTER 10

Conclusion: The Big Picture

You know the old saying, right?

If at first you don't succeed...

Hang with me for a moment during this next story, as there is a UX moral to it, I promise...

When the wonderful woman who today is my wife and I first met, we were working in separate departments at an insurance company in California. I had seen her in the hallways, a woman with beautiful blonde hair, dazzling eyes, and a positive, energetic attitude that was infectious.

Our first true date was a drive along the beach in Malibu. I took her to several of my favorite ocean overlooks, including a nice parking area that sat on the cliffs with an amazing view of the deep blue Pacific Ocean and golden sunset.

Somewhat concerned, I noticed that a couple of times as I drove along the highway she almost seemed to fall asleep. Concerned that this was a signal that maybe she just wasn't that into me, I hoped to jump-start the romance by strolling together along the path at the top of the cliff hand in hand to watch the beautiful sunset.

She seemed to be enjoying the view and my companionship, holding hands with me as we looked out over the breaking surf and the deepening red ball of the sun as it almost touched the sea. I told her I was enjoying the beautiful view and that she was making it even more beautiful. Our bodies were close, our faces near enough that I could gently lean in for a kiss.

I did so, only to find her turning her cheek at the last second, my lips awkwardly grazing the side of her face somewhat like a bug smearing a windshield.

Embarrassed, disappointed, and feeling completely rejected, I pulled back quickly, my mind racing for ideas on how to salvage what was quickly becoming one of the most awkward moments of my life.

I decided on the tried-and-true tactic used by many bad-dates-gone-wrong guys from cave-man days to the present. Retreat.

© W. Craig Tomlin 2018
W. C. Tomlin, *UX Optimization*, https://doi.org/10.1007/978-1-4842-3867-7_10

"Um, hey, you know, it's, um, getting kinda late and I forgot I've got some stuff and things I have to do at home," I lied. "Maybe we should head back," I said in the best nonchalant manner I could muster.

We got into my car, which at that moment I was supremely wishing was a teleportation device to make the trip go instantly, and headed back.

In another of my life's "longest awkward moments award winners" and feeling completely chagrined by what I felt was a highly disinterested woman sitting rather listlessly near me in the passenger seat of my car, we drove back to her house in near silence. It added a bit of salt to my wounds that several more times along the way her head nodded as she almost fell asleep.

As you might be able to guess, I didn't call her after that disastrous date.

A few days went by and by an amazing coincidence, or perhaps a bit of divine guidance, we happened to bump into each other at a bar/restaurant in Santa Monica. I had never been there before, and turns out neither had she.

I tried my best to pretend to not notice her, feeling all over again the shame of our embarrassing sunset date. But she noticed me, smiled, and waved. When I didn't approach her, she came up to me and said a quick hello.

A day later, she called me. She told me she couldn't find her watch and was wondering if maybe it had slipped down the seat of my car.

While speaking with her and cradling the phone awkwardly in the crook of my neck I went to my car and began looking for her watch. I asked her what it looked like and she gave me a highly detailed description. She must really have loved that watch!

I looked high and low in my car, virtually tearing it into pieces trying to find her missing watch. No watch.

As I learned later, she had never lost her watch at all. She was twirling it around her wrist as she was describing it to me on the phone!

While I was searching for the watch, she casually asked me, "So why didn't you call me after our date?"

"Well, to be frank, it seemed to me you weren't that interested. You kept falling asleep and when I tried to kiss you, well, you pulled away," I said, somewhat embarrassed I had to bring up such an awkward moment in my life to of all people - her.

"I'm so sorry," she said quickly. "I really wanted to go out on the date with you, but I had walking pneumonia at the time and was kind of out of it. I didn't want you to kiss me and maybe get it too. But I really wanted to go out with you so I just kind of kept it a secret. I really do like you and I'm sorry if I put you off."

Several emotions swiftly tumbled through my heart and head at hearing this new information. Gladness that she liked me after all. Excitement that perhaps we might continue dating. And a bit of rather annoyed surprise that she had hid something pretty important from me, like the fact that she was a pneumonia factory.

We started dating again, having much better dates than our first attempt. These included successful kisses at the cliffs overlooking the mighty Pacific Ocean with a glowing red sunset. A year later we became engaged. And the happy ending to this story (other than she didn't lose her watch) is we have been married for over 20 years now.

The moral of this story as demonstrated by my wife is if at first you don't succeed, try, try again. There's also perhaps something in there about telling your date if you have a rather serious health condition, but I digress.

You may be wondering what the UX implication is of this little story. It is…

UX design in general and UX research in particular can be very much like a first date gone wrong. No matter how carefully you plan, and no matter how much both parties (you and your users) want things to work, they may not. Things can and do go sideways in unusual and unexpected ways. But don't give up; just try, try again.

So don't be surprised if you find that not everything will come easy or go smoothly as you conduct behavioral UX and usability testing analyses. Sometimes things go sideways. But hang in there, stay persistent, and keep on trying. It gets easier, and eventually you'll be a real pro at this process.

The Big Picture

You've learned a lot of techniques, strategies, tools, and tactics in this book. You're probably anxious to try out many of the tools and techniques you've read about. Yay!

My advice is

Practice makes perfect, things probably won't always work out the way you expect, and that's okay. Learn from what happened, what went right, and what went wrong, and move forward. Take what you've done in past studies and apply changes to what you're doing for future studies.

Another piece of advice I want you to take to heart is

Always test the test.

When it comes to usability testing and related research tests, I cannot begin to tell you the number of times when testing the test has helped me improve the sessions I've conducted. And the interesting thing is, even when I conduct virtually identical usability tests, there's always something new or different that makes each test unique, often causing the need for slight tweaks or adjustments to the test.

So always test your usability or other research test before conducting the actual test. It's pretty easy to do so with the online testing tools that are out there today. It's also easy to do with remote moderated or even in-person moderated usability tests. Friends, family, or coworkers who are not part of your project can be useful testers for your test of the test.

Testing your test may take slightly longer, but the results provided will help to make your test even better, so always test the test.

Most likely as you start creating your own UX research studies you'll find it makes sense to tweak and adjust them to suit your specific needs and style. That's fine! There's many ways to conduct studies and what's right for one person or in one situation may need to be adjusted for another person or in another situation.

I gave you the blueprint, but the house you build is uniquely yours.

Another sage piece of advice is something we all can use, which is

Always be open to new ideas.

That doesn't mean you have to accept every new idea, but you should certainly try to understand new information and evaluate if and how it can help improve what you're doing today.

Case in point, the changes in our industry for some of the newer user experiences people are having with technology. Testing websites and apps for the most part has been pretty standard for many years now. But newer technologies such as the latest user experiences with voice interfaces such as Amazon Echo, AI (artificial intelligence), VR (virtual reality), AR (augmented reality), and MR (mixed reality) are changing the UX research landscape.

The principles, strategies, and tactics I've covered in this book are very helpful for evaluating experiences people have with websites and apps. But some new techniques and tools may be necessary for these newer user experiences that do not use keyboard, mice, or monitors.

Human Computer Interaction has taken on an entirely new and far broader scale with the introduction of these new interfaces. It's my opinion that over time interfaces like keyboards, monitors, and mice will slowly but surely become passé. We are all becoming far more interconnected with technology and intertwined with a broader variety of interfaces. How will all of this impact UX research?

Even now new ways of evaluating and testing usability and user experiences for these newer interfaces are being developed. As time goes by, better and more sophisticated ways of conducting evaluations of the behavioral UX and usability testing data associated with those interfaces will be developed.

How you proceed will be up to you. But keeping an open mind and exploring new ways of evaluating user experiences for devices other than computers will be important. My advice to you is simple…

The future is yours, so don't be afraid to venture outside the norm and outside your comfort zone.

What We've Covered

I hope you feel like you've learned a lot from this book. Let's take a brief look back to summarize everything we have covered so far, chapter by chapter.

Chapter 1: UX Optimization Overview

There are generally two broad types of UX optimizations and two types of optimizers. The first group is quantitative researchers who use data (sometimes big data) to evaluate the "what's happening" information about user experiences. The second group is the qualitative researchers who use usability testing and related techniques to evaluate the "why it's happening" information about user experiences.

What you learned is that by combining both quantitative and qualitative data into a robust UX research study, you can gain a 360-degree view into the "what's happening" and the "why it's happening" data. This view lets you be far more effective at analyzing the user experience and making recommendations for optimizations.

There are four big UX optimization steps when conducting this type of research:

Step 1: Defining Personas
Step 2: Conducting Behavioral UX Data Analysis
Step 3: Conducting UX and Usability Testing
Step 4: Analyzing Results and Making Optimizations

Step 1 is to clearly define who you are optimizing the website for by creating a Persona or Personas, or using existing Personas if they already exist. A Persona is a representation of your most common website visitors who all share the same critical tasks.

Step 2 is to conduct the behavioral UX data analysis to identify the quantitative WHAT behavioral data coming from log file analysis tools such as Google Analytics. You can identify potential areas of the website that may be causing poor critical task performance for your Persona or Personas.

Step 3 is to conduct qualitative WHY UX and usability testing to uncover potential reasons for the poor critical task performance. This testing may align with some of the more common heuristic usability problems that cause website visitors to have difficulty in accomplishing their critical tasks.

Step 4 is where you combine the WHAT behavioral data with the WHY UX and usability testing data. This gives you a much clearer picture of the user experience on the website. With this information, you can now look for patterns and make recommendations for potential optimizations. A/B testing should be conducted on any optimization recommendations to ensure that the recommendations actually do improve the user experience of the website.

Chapter 2: What's a Persona?

A Persona is a representation of the most common users, based on a shared set of critical tasks. Personas are based on field research and direct user observation in their environment (also known as contextual inquiry).

Personas vary broadly, but most generally share attributes including the following:

- They are based on field research and user observation.

- They identify common patterns of behavior.

- They focus on the now, not a potential future state of how things might be.

- They include a picture, name, and brief story to humanize the Persona.

- They describe a problem or task the Persona is trying to solve, typically in a story format.

- Good Personas include the typical environment and/or devices used.

- They include details on the domain expertise of the typical user (how familiar the user is with the language, process flow, and details of usage).

- They call out in prioritized order the top two or three critical tasks the Persona must do to be successful.

There are three general types of Personas:

- Design Personas (UX Personas)

- Marketing Personas (Buyer Personas)

- Proto-Personas (made without field research)

Personas are extremely important to the design and optimization of websites and apps because they help focus teams on the end users and what their needs and mental maps for process flows are. Without Personas, any testing or optimization of a website has the danger of being flawed because the analysis and results may not be focused on helping the end user achieve their goals and critical tasks.

Chapter 3: Types of Personas

There are three general types of Personas:

- Design Personas

- Marketing Personas (also called Buyer Personas)

- Proto-Personas

Design Personas are the most useful for website and app design and optimization. That's because they are created by observing real people in their own environment interacting with tasks similar to the critical tasks being researched for the website or app.

There is no single perfect time for when a Design Persona should be updated. In general, Design Personas should stay consistent with the goals, needs, behaviors, and mental map of the processes most people use when engaging with your website or app.

There are two very broad categories of types of Design Personas that can be considered when evaluating how often to refresh:

- **Consumer Design Personas**: Associated with consumers and individuals using a product or service who are not associated with a firm and who typically have lower domain expertise

- **Business-Based Design Personas**: People who are either employees or business partners with a firm and who may have a higher level of domain expertise

Consumer Design Personas can generally be updated every year to every other year, depending on many variables including the product, industry, types of users, and more.

Business-Based Design Personas can generally be updated every year to every two years, or perhaps longer. This is because in general the internally focused applications and associated critical tasks the Business-Based Design Persona will be using may not be updated as often as consumer-based applications.

Marketing Personas (sometimes referred to as Buyer Personas) are based on demographic, geographic, and other data sources and typically do not include actual observation of real people in their environment. These Personas are often used by marketing, advertising, and product teams when creating communications for a target audience. The Marketing Personas are most useful for identifying needs, pain points, and desires that help marketing teams create the messaging used to attract awareness and engagement of their prospective customers.

Marketing or Buyer Personas, like Design Personas, do not have a set time for updating. Instead, various factors can influence when a Marketing Persona should be updated, including

- New or modified business processes or approaches to work

- Changes in technology or usage of technology

- Shifting awareness, pain points, needs, goals, or priorities

- New approaches, tools, or methods of accomplishing tasks or completing goals

- Adjustments in the mental maps for how processes should be or are completed

Proto-Personas are similar to Design Personas except that Proto-Personas are not created with observation of real people in their own environment. Often, secondary sources of data are used to build Proto-Personas, which enables them to be created quickly, which is a useful feature for agile-based firms that require weekly sprints and design research sessions.

Proto-Personas are easy to update because much of the "creation" of a Proto-Persona is based on indirect data and information vs. the more difficult and time-consuming work required in recruiting and observing real people in their environment.

Some teams update their Proto-Personas with each sprint, using data gathered from users as they interact with the website or app. However, this makes Proto-Personas dangerous because without the direct observation of real users in their own environment it is possible to arrive at the wrong conclusions regarding a website or app design.

Proto-Personas can and should be updated to true Design Personas at the earliest opportunity.

Chapter 4: Why Personas Matter

Personas are critical in UX and in any analysis of UX. Some of the more important reasons focus on helping to bring users and their needs to the forefront of any discussion of functions or features. Personas can also

- Add context to UX behavioral data

- Enable user-centered design

- Aid in recruiting for usability testing

- Decrease scope creep

Personas are critical for determining what data to review as part of a UX behavioral analysis. Personas enable user-centered design by representing users when teams do not have the time or resources to bring real users in for testing but still need to make decisions that impact the user experience.

It's virtually impossible to quickly and accurately recruit for usability testing without having a Persona. Personas are very helpful in helping screeners find testers who match the typical user of a website or app.

Finally, Personas can help teams reduce scope creep by being the litmus test for whether a feature of function is needed. And Personas can help when teams evaluate what a MVP experience should or should not include.

Chapter 5: How to Create a Persona

Most of you probably already have Personas that are ready to use. But for those of you who don't, contextual inquiry is the method used to observe and ask questions of your participants. It means getting out of the office and going to the places where your users are and observing them as they use the systems you are researching. It's also a good idea to review secondary sources of information about users, always remembering they are secondary and cannot replace the contextual inquiry data.

The steps necessary for a successful contextual inquiry include preliminary work of asking yourself questions about the purpose and goal of your sessions. During the session, you want to ask open-ended questions, listen much more than talk, probe, and ask follow-up questions. You want to take copious notes!

After your session you should consolidate your notes and begin looking for patterns.

- What do your participants say repeatedly about their goal or desire? What are they trying to accomplish?

- How does the current system help them accomplish their goal or goals?

- What parts of the current process work well?

- What parts of the current process do not work well?

- What consistent task-flow successes, or failures, are shared among the users?

- What are common pain points?

- What are consistent workarounds to existing problems?

- Is it common for people to have sticky notes with information on or near their computer to help them with their critical task? What's on those stickies?

- Do people frequently rely on cheat sheets or other non-system documents to complete a task? What are those aids?

- Where are there gaps in the existing process? What are those gaps?

As you create your persona you'll want to include several elements of important information, including

1. **Identify Critical Tasks**: What are the top 1-3 critical tasks necessary for the end user to be successful?

2. **Document Environment of Use**: Are there common places, devices, or third-party tools that are consistently used or needed?

3. **Define Domain Expertise**: Is there a common domain expertise, meaning familiarity with the systems, terminology, or processes?

4. **Identify Pain Points**: What are the common pain points shared among users?

5. **Create a Name**: Be sure to be culturally sensitive. Focus on common names that are easy to remember and that can easily be used by your team.

6. **Find a Picture**: As humans, we are visual creatures, so a face and name are important to humanize the Persona.

Creating Persons takes work. If you've not done it before, you may want to practice the entire process a few times with family and friends.

Once you've tried a few sessions, I think you'll agree with me that it is very rewarding to spend time learning more about your users, and that in turn will make your design and development efforts that much better.

Chapter 6: Behavioral UX Data

Behavioral UX data is a very effective way to analyze the quantitative "what's happening" information on your website.

The four broad types of behavioral UX data include

- **Acquisition**: Such as PPC keyword data, etc.

- **Conversion**: Users who "convert" by taking an action, etc.

- **Engagement**: Such as bounce rate, time on page, etc.

- **Technical**: Visits by browser, screen resolution, etc.

There are a variety of sources for behavioral UX data. Many websites use log file analysis tools such as Google Analytics to consolidate the activity on a website into helpful charts, graphs, and reports.

In addition, behavioral UX data can be found in CRM (Customer Relationship Management) systems such as SalesForce or other back-end systems associated with tracking prospective customers or eCommerce activity.

The overwhelming number of reports, graphs, charts, and other quantitative data can seem daunting. But by understanding the critical tasks and needs of the Personas and of the business it is possible to narrow down the glut of information into specific sets of data that will shed light on "what's happening" with the user experience.

Evaluating this data takes a bit of practice but once completed will provide patterns where tasks flows are working well and where other task-flows may not be working as anticipated.

Although UX behavioral data is very helpful, this quantitative "what's happening" data is not enough.

So what's missing?

Although you know WHAT is happening, you still don't know WHY it is happening.

Without this WHY information, any suggestions for optimizations are dangerous because you are forced to guess as to why the behaviors being reported in the behavioral UX data are happening. Rather than guess, there's a much better way to find this all important WHY (or qualitative) data.

You uncover the WHY data by using qualitative UX research and usability testing.

By conducting UX research or usability testing, you can begin to understand the WHY that helps you make sense of the WHAT behavioral UX data you've already documented. Combining both of these data elements (quantitative and qualitative) provides you with a far more accurate set of information with which to make optimization recommendations.

Chapter 7: UX and Usability Testing Analysis

You'll use a variety of UX and usability testing tools to gather qualitative WHY information about website activity.

These tools can include the following:

- **Moderated Usability Test**: Richest source of qualitative data, but resource intensive

- **Unmoderated Usability Test**: Good source of qualitative data, and less resource intensive

- **5-Second Test**: Provides data for how well a page communicates key concepts to visitors

- **Card Sort**: Excellent for evaluating information architecture questions

- **Click Test**: Identifies navigation and taxonomy issues

- **Eye Tracking**: Identifies visual elements that may be helping or hurting in attracting attention

- **Preference Test**: Can add the WHY to preferences among several choices

- **Question Test**: Useful for identifying task-flow and page communication problems

The goal of using these tests is to obtain the all-important WHY qualitative data coming from your website activity. By analyzing this data, you will have a much better picture of what's occurring on your website and possible reasons why it is occurring.

By combining the quantitative WHAT behavioral data you captured earlier with this qualitative WHY data, you now get a much more complete, "360-degree view" of what's happening on the website and why it is happening. This gives you a much better set of data with which to make website optimization recommendations.

Chapter 8: Putting it All Together: Behavioral UX Data Analysis

In analyzing the Behavioral UX data coming from the case study you have now reviewed enough data that you can see some patterns starting to emerge. These patterns include understanding who the typical users are (Personas), pages that experience low or abnormal activity, task flows that have issues in certain places, and certain experiences that indicate where activity or actions seem to be completed at lower than expected volume.

After completing your evaluation of all the behavioral UX data including the above case study examples and more, you'll want to summarize your findings into short, bite-sized pieces of identified patterns.

The goal of your summary of the behavioral UX analysis is to help your client or team understand where you are finding these patterns and the "so what?" of the patterns.

You should include

- Who you are evaluating (the Persona or Personas)

- Irregular behavior flow patterns that hint at pages to evaluate further using usability testing data

- Other patterns of usage that indicate where users are experiencing difficulty or abandoning a critical task

- Recommended next steps, especially around how you use this data to identify what qualitative data to analyze

You might summarize the findings from your detailed investigation of the behavioral UX data into a slide or two in your analysis presentation, which is helpful for discussion and for the next step of the analysis of the usability testing data.

If you're evaluating an eCommerce site, you'll more than likely have one or two additional summary slides focusing on the eCommerce issues found and on any additional patterns of unusual behavior that should be investigated.

The common questions I'm asked at this point are

"Do I show the team the findings so far? Or do I wait for the additional information coming from the usability testing and other qualitative data?"

I believe you already know the answer to these questions, assuming you've read the prior chapters!

The answer is, it's best to wait to present your behavioral UX data analysis because so far all you've uncovered is the "what's happening" quantitative data. You still don't know "why it's happening" qualitative data.

For that, you need to take your quantitative findings and patterns found from the behavioral UX data and use that information to target your attention on specific qualitative usability testing data.

Your goal is to learn "why it's happening."

Chapter 9: Putting it All Together: Usability Testing Data Analysis and Recommendations

This chapter covered several important and useful tools and techniques for evaluating the qualitative "why it's happening" data. You know what qualitative data to focus on based on the results of the quantitative "what's happening" data.

You explored

- The **5-Second Test** and why it helps identify how well the site communicates the three critical elements of who you are, what you do, and why visitors should care

- The importance of **body copy**, **readability**, and **calls to action** to communicate effectively with your website visitors and help them engage with the site

- **Automated eye tracking** and how it can help you identify potential places in a design that are either not attracting enough attention or attracting too much attention

- An understanding of the **eCommerce shopping cart** and the pages and/or fields that may be causing higher abandonment rates including recommendations for optimizations based on social proof and best practices

- **Pricing** and how product descriptions, benefits, and unique attributes compared to the competition can help stimulate purchase consideration

Using the analysis techniques and tools mentioned in this and the prior chapter will greatly help you tell the story of both "what's happening" and "why it's happening." Using this data and the analysis you preformed will provide more accurate and more detailed recommendations for optimizations.

How Do You Get To Carnegie Hall?

The old joke of "How do you get to Carnegie Hall?" "Practice!" is very useful at this point. Now that you've learned what to do, and how to do it using your use case, you now have to practice, and practice, and practice!

The more time you spend conducting behavioral UX and usability research and analysis, the easier it will become. And I think that once you've got the hang of it you'll feel right at home conducting evaluations, analysis, and making recommendations.

Conclusion: The Big Picture

Bravo! I'm very proud of you! If you've come this far in the book, it means you're truly dedicated to learning about and using the UX research practices and methods I've shared with you. You've come a long way indeed!

You've learned a lot about UX research and optimization of websites. You've learned what Personas are, why they are important, and how to create and use Personas. You've learned how to find and analyze quantitative "what's happening" data. You've learned how to find and analyze qualitative "why it's happening" data and you've learned how to combine both into a far more accurate, 360-degree picture of where issues and opportunities may lie for a website.

In addition, you've gone through a case study demonstrating how to put together the patterns to help tell the story of UX issues and opportunities, and how to present them to your clients or internal teams.

Finally, you've learned that the future is wide open in terms of where UX research can go. There are new frontiers to explore in interactive voice, AI, VR, AR, MR, wearables, and many more interactive technologies that will require investigation and testing processes to identify ways to improve them.

But all of this is just the base of the pyramid that is the representation of your knowledge and skills of UX analysis and optimization. There is much more to learn, many more tools and techniques to try, and an almost infinite number of experiences to evaluate. Putting the rest of the blocks on your pyramid is up to you. I hope if you've not already done so that you will go out today and start trying your hand at some of the many ways to find, analyse, and optimize user experiences.

And should you have a moment or two, I would really appreciate hearing from you on how you're coming along. And don't be afraid to share with me any ideas you have on what you liked about this book, or where you think I could make improvements. I'm always interested in improving user experiences, even with my own book! Just drop me a note on my UsefulUsability Facebook page or send a tweet to my @ctomlin Twitter feed. I'll be very grateful for any thoughts or feedback you have!

Best of luck to you, now and always!

Index

© W. Craig Tomlin 2018
W. C. Tomlin, *UX Optimization*, https://doi.org/10.1007/978-1-4842-3867-7